Series/Number 07-106

APPLIED LOGISTIC
REGRESSION ANALYSIS

SCOTT MENARD
Institute of Behavioral Science
University of Colorado, Boulder

SAGE PUBLICATIONS
International Educational and Professional Publisher
Thousand Oaks   London   New Delhi

For information address:

SAGE Publications, Inc.
2455 Teller Road
Thousand Oaks, California 91320
E-mail: order@sagepub.com

SAGE Publications Ltd.
6 Bonhill Street
London EC2A 4PU
United Kingdom

SAGE Publications India Pvt. Ltd.
M-32 Market
Greater Kailash I
New Delhi 110 048 India

Printed in the United States of America

Library of Congress Cataloging-in-Publication Data

Menard, Scott W.
    Applied logistic regression analysis / Scott Menard.
        p.   cm. — (Sage university papers series. Quantitative
    applications in the social sciences; no. 07-106)
    Includes bibliographical references (pp. 96-97).
    ISBN 0-8039-5757-2 (pbk.: acid-free)
    1. Regression analysis. 2. Logistic distribution.   I. Title.
    II. Series.
    QA278.2.M46   1995
    519.5'36—dc20   95-9071

95 96 97 98 99 10 9 8 7 6 5 4 3 2 1

Sage Project Editor: Susan McElroy

When citing a university paper, please use the proper form. Remember to cite the current Sage University Paper series title and include the paper number. One of the following formats can be adapted (depending on the style manual used):

(1) MENARD, S. (1995) *Applied Logistic Regression Analysis.* Sage University Paper series on Quantitative Applications in the Social Sciences, 07-106. Thousand Oaks, CA: Sage.

OR

(2) Menard, S. (1995). *Applied logistic regression analysis* (Sage University Paper series on Quantitative Applications in the Social Sciences, series no. 07-106). Thousand Oaks, CA: Sage.

CONTENTS

SERIES EDITOR'S INTRODUCTION

Regression, the most widely used nonexperimental data analysis technique, receives further extension, in the move from ordinary least squares (OLS) regression to logistic regression. To facilitate understanding of the latter, Dr. Menard begins chapters with OLS regression principles, which are then appropriately modified or paralleled for logistic regression. Chapter 1 introduces the problems a dichotomous variable ($Y = 1, 0$) presents for OLS assumptions; for example, heteroscedasticity, nonnormal error term, nonlinearity, and predicted probabilities beyond 1.0. He shows how these assumption violations can be overcome with a logit dependent variable; that is, a dependent variable that is the natural log of the odds of Y occurring or not. Maximum likelihood estimation is then in order, rendering more efficient estimates than OLS.

Chapter 2 reviews summary statistics from OLS, then offers logistic counterparts. Analogous to the R^2, for example, is an R_L^2 or a pseudo R^2. Also, several measures of proportional reduction in prediction error are given. As of this writing, many of these measures are not available in the usual software packages. However, Dr. Menard explains that most can be calculated manually. Chapter 3 focuses on interpretation of the coefficients. With OLS, of course, a unit change in X brings, on average, a slope value (b) change in Y. What can be said about changes in the coefficients of logistic regression? One important interpretation comes from translating logit coefficients into probability statements, something that Dr. Menard lucidly explicates. For instance, in a telling illustration, he works through how respondents in a National Youth Survey have a .998 probability of using marijuana if they have a maximum score on a particular X, but only a .097 probability if they have a minimum score on that X. He goes on to discuss standardized coefficients in logistic regression. Perhaps surprisingly, it is revealed that they have a similar interpretation to standardized coefficients in OLS regression. Further, he develops the "odds ratio" interpretation but cautions that it does not really replace the standardized logistic coefficient when one wishes to evaluate the relative strengths of the independent variables. Chapter 4 looks at regression diagnostics but in a logistic context. Those who have studied ordinary regression diagnostics will find the list of problems familiar: omitted relevant variables, included irrelevant variables, nonlinearity, nonadditivity, collinearity, and outliers. Chapter 5 explores logistic regression when the dependent variable is polytomous, rather than dichotomous. Methodological work, as well as published research examples, emphasizes the simple case of the two-category dependent variable, but, obviously, the situation of a dependent variable with three or more categories is common and deserves more application. In the polytomous case, there is a reference category, to which the probabilities of placement in the other categories are compared. Supposing M categories, $M - 1$ equations must be estimated. In the author's example, the dependent variable is drug use and the reference category is drug nonusers, leading to estimation of three logit functions: one for alcohol users versus drug nonusers, another for marijuana users versus drug nonusers, and a final one for illicit drug users versus drug nonusers. Dr. Menard observes that when the dependent variable is actually measured on an ordinal scale, other possibilities exist besides logistic regression, and might even be preferred. (Here OLS returns as one of the possibilities.) He emphasizes, however, that the choice of technique requires thought: "Mechanical application of options available in existing software packages is not recommended." For those who have mastered applied regression (perhaps from *Applied Regression: An Introduction*, No. 22, this series), this tidy monograph on applied logistic regression is a logical next step up the learning ladder.

—*Michael S. Lewis-Beck*
Series Editor

APPLIED LOGISTIC REGRESSION ANALYSIS

SCOTT MENARD
Institute of Behavioral Science
University of Colorado, Boulder

1. LINEAR REGRESSION AND THE LOGISTIC REGRESSION MODEL

In linear regression analysis, it is possible to test whether two variables are linearly related, and to calculate the strength of the linear relationship, if the relationship between the variables can be described by an equation of the form $Y = \alpha + \beta X$, where Y is the variable being predicted (the dependent, criterion, outcome, or endogenous variable), X is a variable whose values are being used to predict Y (the independent, exogenous, or predictor variable),[1] and α and β are population parameters to be estimated. The parameter α, called the *intercept*, represents the value of Y when X is zero. The parameter β represents the change in Y associated with a one-unit increase in X, or the *slope* of the line that provides the best linear estimate of Y from X. In *multiple regression*, there are several predictor variables. If k is the number of independent variables, the equation becomes $Y = \alpha + \beta_1 X_1 + \beta_2 X_2 + \ldots + \beta_k X_k$ and $\beta_1, \beta_2, \ldots, \beta_k$ are called *partial* slope coefficients, reflecting the fact that any one of the k predictor variables X_1, $X_2, \ldots, X_k$ provides only a partial explanation or prediction for the value of Y. The equation is sometimes written in a form that explicitly recognizes that prediction of Y from X may be imprecise: $Y = \alpha + \beta X + \epsilon$ or, for several predictors, $Y = \alpha + \beta_1 X_1 + \beta_2 X_2 + \ldots + \beta_k X_k + \epsilon$, where ϵ is the error term, a random variable representing the error in predicting Y from X. For an individual case j, $Y_j = \alpha_j + \beta X_j + \epsilon_j$ or $Y_j = \alpha_j + \beta_1 X_{1j} + \beta_2 X_{2j} + \ldots + \beta_k X_{kj} + \epsilon_j$, and the subscript j indicates that the equation is predicting values for specific cases, indexed by j ($j = 1$ for the first case, $j = 2$ for the second case, etc.). Y_j, X_{1j}, X_{kj}, and so on, refer to specific values of the dependent and independent variables. This last equation is used to calculate the value of Y for a particular case, j, rather than describing the relationship among the variables for all of the cases in the sample or the population.

1

Estimates of the intercept, α, and the regression coefficients, β (or β_1, $\beta_2, \ldots, \beta_k$), are obtained mathematically using the method of ordinary least squares (OLS) estimation, which is discussed in many introductory statistics texts (e.g., Agresti & Finlay, 1986; Bohrnstedt & Knoke, 1988). These estimates produce the equation $\hat{Y} = a + bX$ or, in the case of several predictors, $\hat{Y} = a + b_1 X_1 + b_2 X_2 + \ldots + b_k X_k$, where $\hat{Y}$ is the value of Y predicted by the linear regression equation, a is the OLS estimate of the intercept α, and b (or b_1, b_2, $\ldots$, b_k) is the OLS estimate for the slope β (or the partial slopes β_1, β_2, $\ldots$, β_k). Residuals for each case, e_j, are equal to $(Y_j - \hat{Y}_j)$, where $\hat{Y}_j$ is the estimated value of Y for case j. For bivariate regression they can be visually or geometrically represented by the vertical distance between each point in a bivariate scatterplot and the regression line. For multiple regression, visual representation is much more difficult because it requires several dimensions.

An example of a bivariate regression model is given in Figure 1.1. In Part A of Figure 1.1, the dependent variable is FRQMRJ5, the annual FReQuency of self-reported MaRiJuana use ("How many times in the last year have you smoked marijuana?"), and the independent variable is EDF5, an index of Exposure to Delinquent Friends, for 16-year-old respondents interviewed in 1980 in the fifth wave of a national household survey.[2] The exposure to delinquent friends scale is the sum of the answers to eight questions about how many of the respondent's friends are involved in different types of delinquent behavior (theft, assault, drug use). The responses to individual items range from 1 (none of my friends) to 5 (all of my friends), resulting in a possible range from 8 to 40 for EDF5. From Part A of Figure 1.1, there appears to be a positive relationship between exposure to delinquent friends and marijuana use, described by the equation

$$(FRQMRJ5) = -49.2 + 6.2\,(EDF5).$$

In other words, for every one-unit increase in the index of exposure to delinquent friends, frequency of marijuana use increases by about six times per year, or about once every 2 months. The coefficient of variation, or R^2, indicates how much better we can predict the dependent variable from the independent variable than we could predict the dependent variable without information about the independent variable. Without information about the independent variable, we would use the mean frequency of marijuana use as our prediction for all respondents. Knowing the value of exposure to delinquent friends, however, we can base our prediction on the value of

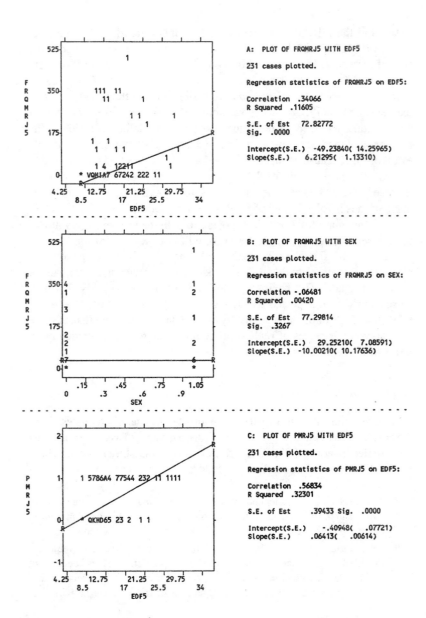

Figure 1.1. Bivariate Regression Plots

EDF5 and the relationship, represented by the regression equation, between FRQMRJ5 and EDF5. Using the regression equation reduces the sum of the squared errors of prediction, $\sum e_j^2 = \sum(\hat{Y}_j - Y_j)^2$, by $R^2 = .116$, or about 12%.

It is necessary in interpreting the results to consider the actual values of the dependent and independent variables. The intercept indicates that for an individual with zero as the value of exposure to delinquent friends, the frequency of marijuana use would be negative. This seemingly impossible result occurs because exposure, as noted above, is measured on a scale that ranges from a minimum of 8 (no exposure at all; none of one's friends involved in any of 8 delinquent activities) to 40 (extensive exposure; all of one's friends involved in all of 8 delinquent activities). Thus, for individuals with the minimum possible exposure to delinquent friends (a value of 8, representing no exposure), the expected frequency of marijuana use is $-49.2 + 6.2(8) = 0.4$, close to zero but indicating that even among individuals with no exposure to delinquency, there are some who use marijuana at least occasionally. The maximum value of EDF5 in this sample is 29, which corresponds to an expected frequency of marijuana use equal to $-49.2 + 6.2(29) = 130.6$, or use approximately every 3 days. This result makes sense substantively, in terms of real-world behavior, as well as statistically, in terms of the regression equation.

Regression Assumptions

In order to use the OLS method to estimate and make inferences about the coefficients in linear regression analysis, a number of assumptions must be satisfied (Berry, 1993; Berry & Feldman, 1985; Lewis-Beck, 1980, pp. 26-47). Specific assumptions include the following:

1. *Measurement:* All independent variables are interval, ratio, or dichotomous, and the dependent variable is continuous, unbounded, and measured on an interval or ratio scale. All variables are measured without error.[3]
2. *Specification:* (a) All relevant predictors of the dependent variable are included in the analysis, (b) no irrelevant predictors of the dependent variable are included in the analysis, and (c) the form of the relationship (allowing for transformations of dependent or independent variables) is linear.
3. *Expected value of error:* The expected value of the error, ϵ, is zero.
4. *Homoscedasticity:* The variance of the error term, ϵ, is the same, or constant, for all values of the independent variables.

5. *Normality of errors:* The errors are normally distributed for each set of values of the independent variables.

6. *No autocorrelation:* There is no correlation among the error terms produced by different values of the independent variables. Mathematically, $E(\epsilon_i, \epsilon_j) = 0$.

7. *No correlation between the error terms and the independent variables:* The error terms are uncorrelated with the independent variables. Mathematically $E(\epsilon_j, X_j) = 0$.

8. *Absence of perfect multicollinearity:* For multiple regression, none of the independent variables is a perfect linear combination of the other independent variables; mathematically, for any i, $R_i^2 < 1$, where R_i^2 is the variance in the independent variable X_i that is explained by all of the other independent variables $X_1, X_2, \ldots, X_{i-1}, X_{i+1}, \ldots, X_k$. If there is only one predictor, multicollinearity is not an issue.

Violations of the Measurement Assumption:
Dichotomous Variables in Linear Regression

The linear regression model can easily be extended to accommodate dichotomous predictors, including sets of dummy variables (Berry & Feldman, 1985, pp. 64-75; Hardy, 1993; Lewis-Beck, 1980, pp. 66-71). An example is presented in Part B of Figure 1.1. Here, the dependent variable is again self-reported annual frequency of marijuana use, but the independent variable this time is sex or gender (coded 0 = female, 1 = male). The regression equation is

$$FRQMRJ5 = 29.3 - 10.0(SEX).$$

The resulting diagram consists of two columns of values for frequency of marijuana use, one representing females and one representing males. With a dichotomous predictor, coded 0-1, the intercept and the slope have a special interpretation. It is still true that the intercept is the predicted value of the dependent variable when the independent variable is zero (substantively, when the respondent is female), but with only two groups the intercept now is the *mean* frequency of marijuana use for the group coded as zero (females). The slope is still the change in the dependent variable associated with a one-unit change in the independent variable, but with only two categories; that value becomes the *difference in the means* between the first (female) and second (male) groups. The sum of the slope and the intercept ($29.3 - 10.0 = 19.3$) is therefore the mean frequency of marijuana use for the second group (males). As indicated in Part B of

Figure 1.1, then, females report a higher (yes, higher) frequency of marijuana use than males, but the difference is not statistically significant (as indicated by sig. = .3267). In Part B of Figure 1.1, the regression line is simply the line that connects the mean frequency of marijuana use for females and the mean frequency of marijuana use for males, that is, the *conditional means*[4] of marijuana use for females and males, respectively. The predicted values of Y over the observed range of X lie well within the observed (and possible) values of Y. Again, the results make substantive as well as statistical sense.

When the dependent variable is dichotomous, the interpretation of the regression equation is not as straightforward. In Part C of Figure 1.1, the independent variable is again exposure to delinquent friends, but now the dependent variable is the prevalence of marijuana use: whether (yes = 1 or no = 0) the individual used marijuana at all during the past year. In Part C of Figure 1.1, with a dichotomous dependent variable, there are two rows (rather than columns, as in Part B). The linear regression model with a dichotomous dependent variable, coded 0-1, is called a *linear probability model* (Agresti, 1990, p. 84; Aldrich & Nelson, 1984). The equation, from Part C of Figure 1.1, is

$$PMRJ5 = -.41 + .064\,(EDF5).$$

When there is a dichotomous dependent variable, the mean of the variable is a function of the probability[5] that a case will fall into the higher of the two categories for the variable. Coding the values of the variable as 0 and 1 produces the result that the mean of the variable is the proportion of cases in the higher of the two categories of the variable, and the predicted value of the dependent variable (the conditional mean, given the value of X and the assumption that X and Y are linearly related) can be interpreted as the *predicted probability* that a case falls into the higher of the two categories on the dependent variable, given its value on the independent variable. Ideally, we would like the predicted probability to lie between 0 and 1, because a probability cannot be less than 0 or more than 1.

As is evident from Part C of Figure 1.1, the predicted values for the dependent variable may be higher or lower than the possible values of the dependent variable. For the minimum value of EDF5 (EDF5 = 8), the predicted prevalence of marijuana use (i.e., the predicted probability of marijuana use) is $-.41 + .064(8) = .10$, a reasonable result; but for the maximum value of EDF5 (EDF5 = 29), the predicted probability of marijuana use becomes $-.41 + .064(29) = 1.45$, or an impossibly high probability of about 1½. In addition, the variability of the residuals will depend on the size of the independent variable (Aldrich & Nelson, 1984, p. 13; Schroeder, Sjoquist, & Stephan, 1986, pp. 79-80). This condition, called *heteroscedasticity*, implies that the estimates for the regression coefficients, although unbiased (not systematically too high or too low), will not be the best estimates in the sense of having a small standard error. There is also a systematic pattern to the values of the residuals, depending on the value of X. For values of X greater than 23.5 in Part C of Figure 1.1, all of the residuals will be negative because $\hat{Y}_j$ will be greater than Y_j (because for X greater than 23.5, $\hat{Y}_j$ is greater than 1 but Y_j is less than or equal to 1). Also, residuals will not be normally distributed (Schroeder et al., 1986, p. 80) and sampling variances will not be correctly estimated (Aldrich & Nelson, 1984, pp. 13-14); therefore the results of hypothesis testing or construction of confidence intervals for the regression coefficients will not be valid.

Nonlinearity, Conditional Means,
and Conditional Probabilities

For continuous dependent variables, the regression estimate of Y, $\hat{Y}$, may be thought of as an *estimate* of the conditional mean of Y for a particular value of X, *given that the relationship between X and Y is linear.* In bivariate regression, for continuous independent variables, the estimated value of Y may not be exactly equal to the mean value of Y for those cases, because the conditional means of Y for different values of X may not lie exactly on a straight line. For a dichotomous predictor variable, the regression line will pass exactly through the conditional means of Y for each of the two categories of X. If the conditional means of FRQMRJ5 are plotted against the dichotomous predictor, SEX, the plot consists of two points (remember, the cases are aggregated by the value of the independent variable), the conditional means of Y for males and females. The simplest, most parsimonious description of this plot is a straight line between the two conditional means, and the linear regression model appears to work well.

The inherent nonlinearity of relationships involving dichotomous dependent variables is illustrated in Figure 1.2. Here, the observed conditional mean of PMRJ5, the prevalence of marijuana use, is plotted for each value of the independent variable EDF5. The observed conditional mean is symbolized by the letter C. Because PMRJ5 is coded as either 0 or 1, the conditional means represent averages of 0s and 1s, and are interpretable as conditional probabilities. Figure 1.2 is therefore a plot of probabilities that

8

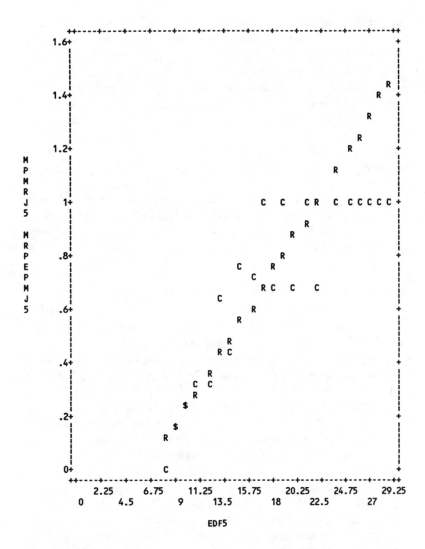

Figure 1.2. Conditional Probabilities Observed (C) and Predicted by Linear Regression (R)

C: observed mean prevalence of marijuana use (MPMRJ5) with EDF5 (exposure to delinquent friends)
R: linear regression prediction of prevalence of marijuana use (MRPEPMJ5) with EDF5
$: multiple occurrence (linear regression prediction and observed value coincide)
21 cases

PMRJ5 = 1 for different values of EDF5. All of the observed values of Y lie between the two vertical lines at 0 and 1, respectively, in Figure 1.2. Predicted probabilities, however, can, in principle, be infinitely large or small, if we use the linear probability model.

The plot of observed conditional probabilities (C) in Figure 1.2 is overlaid with the plot of predicted conditional probabilities based on the regression equation (R) in Part C of Figure 1.1. For values of EDF5 greater than 23.5, the observed value of the conditional mean prevalence of marijuana use stops increasing and levels off at PMRJ5 = 1. The predicted values from the regression equation, however, continue to increase past the value of 1 for PMRJ5, to a maximum of 1.45, and the error of prediction increases as EDF5 increases from 23.5 to its maximum of 29.

Two points need to be made about Figure 1.2. First, although a linear model appears to be potentially appropriate for a continuous dependent variable, regardless of whether the independent variables are continuous or dichotomous, it is evident that a nonlinear model is better suited to the analysis of the dichotomous variable PMRJ5. In general, for very high values of X (or very low values, if the relationship is negative), the conditional probability that $Y = 1$ will be so close to 1 that it should change little with further increases in X. This is the situation illustrated in Figure 1.2. It is also the case that for very low values of X (or very high values if the relationship is negative) the conditional probability that $Y = 1$ will be so close to zero that it should change little with further decreases in X. The curve representing the relationship between X and Y should therefore be very shallow, with a slope close to zero, for very high and very low values of X, if X can, in principle, become indefinitely large or indefinitely small. If X and Y are related, then between the very high and very low values of X the slope of the curve will be steeper, significantly different from zero. The general pattern is that of an "S-curve" as depicted in Figure 1.3.

Second, for prevalence data, the observed conditional mean of Y is equal to the observed conditional probability that $Y = 1$, and the predicted value of Y is equal to the predicted conditional probability that $Y = 1$. The actual values used to identify the two categories of Y are arbitrary, a matter of convenience. They may be 0 and 1, for example, or 2 and 3 (in which case the predicted values of Y are equal to 2 plus the conditional probability that $Y = 3$, still a function of the conditional probability that Y has the higher of its two values for a given value of X). What is substantively important is not the numerical value of Y, but the probability that Y has one or another of its two possible values, and the extent to which that probability depends on one or more independent variables.

10

P
r
e
d
i
c
t
e
d

v
a
l
u
e

o
f

Y

1

0

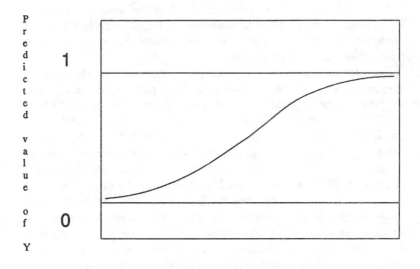

Figure 1.3. Logistic Curve Model for a Dichotomous Dependent Variable

The distinction between the arbitrary numerical value of Y, upon which OLS bases its parameter estimates, and the probability that Y has one or the other of its two possible values is problematic for OLS linear regression and leads us to a consideration of alternative methods for estimating parameters to describe the relationship between X and Y. First, however, we address the issue of nonlinearity. For continuous independent and dependent variables, the presence of nonlinearity in the relationship between X and Y may sometimes be addressed by the use of nonlinear transformations of dependent or independent variables (Berry & Feldman, 1985). Similar techniques play a part in estimating relationships involving dichotomous dependent variables.

Nonlinear Relationships and Variable Transformations

When a relationship appears to be nonlinear, it is possible to transform either the dependent variable or one or more of the independent variables so that the *substantive* relationship remains nonlinear but the *form* of the relationship is linear, and can therefore be analyzed using OLS estimation. Another way of saying that a relationship is substantively nonlinear but formally linear is to say that the relationship is *nonlinear in terms of its variables but linear in terms of its parameters* (Berry & Feldman, 1985, p. 53). Examples of variable transformations to achieve a linear form for the relationship are given in Berry and Feldman (1985, pp. 55-72) and Lewis-Beck (1980, pp. 43-47).

In Figure 1.2, there was some evidence of nonlinearity in the relationship between frequency of marijuana use and exposure to delinquent friends. One possible transformation that could be used to model this nonlinearity is a logarithmic transformation[6] of the dependent variable, FRQMRJ5. This is done by adding 1 to FRQMRJ5 and then taking the natural logarithm. (Adding 1 is necessary to avoid taking the natural logarithm of zero, which is undefined.) The regression equation then has the form $\ln(Y + 1) = \alpha + \beta X$, or equivalently, $(Y + 1) = e^{\alpha + \beta X}$, or $Y = e^{\alpha + \beta X} - 1$, where e = 2.72 is the base of the natural logarithm. Specifically, for prevalence of marijuana use and exposure to delinquent friends,

$$\ln(\text{FRQMRJ5} + 1) = -1.7 + .23(\text{EDF5}); \qquad R^2 = .32.$$

Comparing the results of the model using the logarithmic transformation with the untransformed model in Part A of Figure 1.1, it is evident that the slope is still positive but the numerical value of the slope has changed (because the units in which the dependent variable is measured have changed from frequency to logged frequency). The coefficient of determination for the transformed equation is also larger (.32 instead of .12), reflecting a better fit of the linear regression model when the dependent variable is transformed. This is evidence (not conclusive proof, just evidence) that the relationship between frequency of marijuana use and exposure to delinquent friends is substantively nonlinear. A similar result occurs for the relationship between the dichotomous predictor, SEX, and frequency of marijuana use. With the logarithmic transformation of the dependent variable, the explained variance increases (from a puny .004 to an unimpressive .028), and the relationship between gender and frequency of marijuana use is statistically significant ($p = .011$) in the transformed equation. It appears that the relationship between frequency of marijuana use and both of the predictors considered so far is substantively nonlinear, but we are still able to use a formally linear model to describe those relationships, and we are still able to use OLS to estimate the parameters of the model.

Probabilities, Odds, Odds Ratios, and the Logit Transformation for Dichotomous Dependent Variables

As noted earlier, for a dichotomous dependent variable, the numerical value of the variable is arbitrary, a matter of convenience, and is not intrinsically interesting. What is intrinsically interesting is whether the classification of cases into one or the other of the categories of the dependent variable can be predicted by the independent variable. Instead of trying to predict the arbitrary value associated with a category, then, it may be useful to reconceptualize the problem as trying to predict the probability that a case will be classified into one as opposed to the other of the two categories of the dependent variable. Because the probability of being classified into the first or lower-valued category, $P(Y = 0)$, is equal to 1 minus the probability of being classified into the second or higher-valued category, $P(Y = 1)$, if we know one probability we know the other: $P(Y = 0) = 1 - P(Y = 1)$.

We could try to model the probability that $Y = 1$ as $P(Y = 1) = \alpha + \beta X$, but we would again run into the problem that although observed values of $P(Y = 1)$ must lie between 0 and 1, predicted values may be less than 0 or greater than 1. A step toward solving this problem would be to replace the *probability* that $Y = 1$ with the *odds* that $Y = 1$. The *odds* that $Y = 1$, written "Odds$(Y = 1)$," is the ratio of the probability that $Y = 1$ to the probability that $Y \neq 1$. Odds$(Y = 1)$ is equal to $P(Y = 1)/[1 - P(Y = 1)]$. Unlike $P(Y = 1)$, the odds has no fixed maximum value, but like the probability, it has a minimum value of zero.

One further transformation of the odds produces a variable that varies, in principle, from negative infinity to positive infinity. The *natural logarithm of the odds*, $\ln\{P(Y = 1)/[1 - P(Y = 1)]\}$, is called the *logit* of Y. The logit of Y, written "logit(Y)," becomes negative and increasingly large in absolute value as the odds decrease from 1 to 0, and becomes increasingly large in the positive direction as the odds increase from 1 to infinity. If we use the natural logarithm of the odds that $Y = 1$ as our dependent variable, we no longer face the problem that the estimated probability may exceed the maximum or minimum possible values for the probability. The equation for the relationship between the dependent variable and the independent variables then becomes

$$\text{logit}(Y) = \alpha + \beta_1 X_1 + \beta_2 X_2 + \ldots + \beta_k X_k. \tag{1.1}$$

We can convert logit(Y) back to the odds by *exponentiation*, calculating Odds$(Y = 1) = e^{\text{logit}(Y)}$. This results in the equation

$$\text{Odds}(Y=1) = e^{\ln[\text{Odds}(Y=1)]} = e^{\alpha + \beta_1 X_1 + \beta_2 X_2 + \ldots + \beta_k X_k} \qquad (1.2)$$

and a change of one unit in X multiplies the odds by e^β. We can then convert the odds back to the probability that ($Y = 1$) by the formula $P(Y = 1) = \text{Odds}(Y = 1)/[1 + \text{Odds}(Y = 1)]$. This produces the equation

$$P(Y=1) = \frac{e^{\alpha + \beta_1 X_1 + \beta_2 X_2 + \ldots + \beta_k X_k}}{1 + e^{\alpha + \beta_1 X_1 + \beta_2 X_2 + \ldots + \beta_k X_k}} \qquad (1.3)$$

It is important to understand that the probability, the odds, and the logit are three different ways of expressing exactly the same thing. Of the three measures, the probability or the odds is probably the most easily understood. Mathematically, however, the logit form of the probability is the one that best helps us to analyze dichotomous dependent variables. Just as we took the natural logarithm of the continuous dependent variable, frequency of marijuana use, to correct for the nonlinearity in the relationship between frequency of marijuana use and exposure to delinquent friends, we can also take the logit of the dichotomous dependent variable, prevalence of marijuana use, to correct for the nonlinearity in the relationship between prevalence of marijuana use and exposure to delinquent friends.

For any given case, $\text{logit}(Y) = \pm\infty$. This ensures that the probabilities estimated for the probability form of the model (Equation 1.3) will not be less than 0 or greater than 1, but it also means that because the linear form of the model (Equation 1.1) has infinitely large or small values of the dependent variable, OLS cannot be used to estimate the parameters. Instead, *maximum likelihood* techniques are used to maximize the value of a function, the *log-likelihood* function, which indicates how likely it is to obtain the observed values of Y, given the values of the independent variables and parameters, $\alpha, \beta_1, \ldots, \beta_k$. Unlike OLS, which is able to solve directly for the parameters, for the logistic regression model the solution is found by beginning with a tentative solution, revising it slightly to see if it can be improved, and repeating the process until the change in the likelihood function from one step of the process to another is negligible. This process of repeated estimation, testing, and re-estimation is called *iteration*, and the process of obtaining a solution from repeated estimation is called an *iterative* process. When the change in the likelihood function from one step to another becomes negligible, the solution is said to *converge*. All of this is done by means of computer-implemented numerical algorithms designed to search for and identify the best set of parameters to maximize the log-likelihood function.[7] When the assumptions of OLS regression are met, however, *the OLS estimates for the linear regression coefficients are identical to the estimates one would obtain using maximum likelihood estimation* (Eliason, 1993, pp. 13-18). OLS estimation is in this sense a special case of maximum likelihood estimation, one in which it is possible to calculate a solution directly, without iteration.

Logistic Regression: A First Look

Part C of Figure 1.1 showed the results of an OLS linear regression analysis of the relationship between prevalence of marijuana use (PMRJ5) and exposure to delinquent friends (EDF5). Figure 1.4 presents the output from a bivariate logistic regression with the same two variables. This is output from SPSS LOGISTIC REGRESSION, with some deletions but nothing added. The equation for the logit of the prevalence of marijuana use from Figure 1.4 is

$$\text{logit(PMRJ5)} = -5.4871 + .4068\,(\text{EDF5}).$$

There are several other statistics presented in Figure 1.4, which will be discussed in the pages to follow. For the moment, however, note that the presentation of logistic regression results includes (a) some summary statistics for the goodness of fit of the model (the chi-square statistics); (b) a comparison of observed and predicted values (or classification) of cases according to whether they do (yes) or do not (no) report using marijuana; (c) the estimated parameters (*B*) of the logistic regression equation, along with other statistics associated with those parameters; and (d) a plot of the observed (yes = 1 or no = 0) and predicted probabilities of "membership" of being marijuana users.

Figure 1.5 plots the predicted and observed conditional probabilities (or equivalently the conditional means) for the logistic regression equation. The observed conditional probabilities are represented by the letter "C" and the predicted conditional probabilities by the letter "L" for logistic regression. In Figure 1.2, the predicted probabilities from linear regression analysis represented a straight line, and, for values of EDF5 greater than 23.5, the predicted conditional probabilities of being a marijuana user were greater than 1. The observed conditional probabilities, unlike the predicted conditional probabilities, leveled off at 1. In Figure 1.5, the conditional probabilities predicted by logistic regression analysis all lie between 0 and 1, and the pattern of the predicted probabilities follows the curve suggested

```
logistic regression var=pmrj5 with edf5/method=enter edf5/classplot/save=pred(lpepmrj5).

Dependent Variable..  PMRJ5

Beginning Block Number  0.  Initial Log Likelihood Function -2 Log Likelihood  299.30563

* Constant is included in the model.
Variable(s) Entered on Step Number 1..    EDF5

Estimation terminated at iteration number 4 because Log Likelihood decreased by less than .01 percent.

                       Chi-Square   df  Significance
-2 Log Likelihood       213.947    229    .7543
Model Chi-Square         85.359      1    .0000
Improvement              85.359      1    .0000
Goodness of Fit         236.483    229    .3531

Classification Table for PMRJ5
                  Predicted
                                  Percent Correct
               0 ¦  1
Observed    +-------+-------+
         0  ¦ 136  ¦  14  ¦   90.67%
            +-------+-------+
         1  ¦  37  ¦  44  ¦   54.32%
            +-------+-------+
                 Overall  77.92%

-------------------- Variables in the Equation ----------------------

Variable        B      S.E.     Wald    df    Sig     R    Exp(B)

EDF5          .4068   .0584   48.5460    1   .0000  .3944  1.5020
Constant    -5.4871   .7100   59.7315    1   .0000

              Observed Groups and Predicted Probabilities
       80 +                                                          +
        ¦                                                            ¦
   F    ¦                                                            ¦
   R  60 +                                                           +
   E    ¦   0                                                        ¦
   Q    ¦   0                                                        ¦
   U    ¦   0                                                        ¦
   E  40 +   0                                                       +
   N    ¦   0                                                        ¦
   C    ¦   0  1                                                     ¦
   Y    ¦   0  0  1   1                                              ¦
      20 +   0  0  0   1    1                                        +
        ¦   0  0  0   0    0    1                                    ¦
        ¦   0  0  0   0    0    1    1   1   1                       ¦
        ¦   0  0  0   0    0    0    0   1   0   1  1  1 1111¦
Predicted -------------+--------------+--------------+---------------
    Prob:  0          .25            .5            .75            1
    Group: 00000000000000000000000000000011111111111111111111111111111111
           Predicted Probability is of Membership for 1.00

           Symbols: 0 - .00
                    1 - 1.00

           Each Symbol Represents 5 Cases.

1 new variables have been created.
Name        Contents
LPEPMRJ5    Predicted Value
```

Figure 1.4. Bivariate Logistic Regression for Prevalence of Marijuana Use

by the observed conditional probabilities, a curve similar to the right half of the curve in Figure 1.3. Just from looking at the pattern, there appears

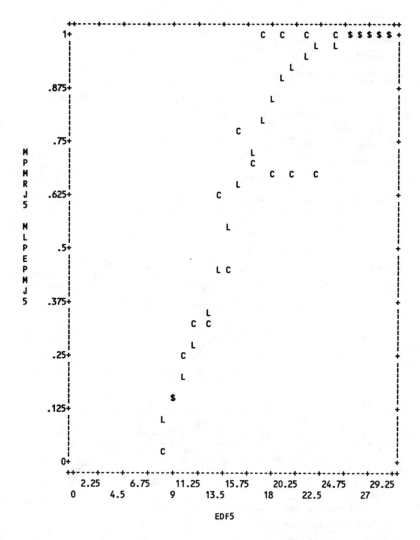

Figure 1.5. Conditional Probabilities Observed (C) and Predicted by Logistic Regression (L)

C: mean prevalence of marijuana use (MPMRJ5) with EDF5 (exposure to delinquent friends)
L: logistic regression prediction of prevalence of marijuana use (MLPEPMRJ5) with EDF5
$: multiple occurrence (logistic regression prediction and observed value coincide)
21 cases

to be a closer correspondence between the observed and predicted conditional means when logistic regression is used to predict the dependent variable.

2. SUMMARY STATISTICS FOR
EVALUATING THE LOGISTIC REGRESSION MODEL

When we evaluate a linear regression model, the evaluation typically has three parts. First, how well does the overall model work? Can we be confident that there is a relationship between all of the independent variables, taken together, and the dependent variable, above and beyond what we might expect as a coincidence, attributable to random variation in the sample we analyze? If there is a relationship, how strong is it? Second, if the overall model works well, how important is each of the independent variables? Is the relationship between any of the variables attributable to random sample variation? If not, how much does each independent variable contribute to our ability to predict the dependent variable? Which variables are stronger or weaker, better or worse, predictors of the dependent variable? Third and finally, does the form of the model appear to be correct? Do the assumptions of the model appear to be satisfied? In this chapter, we deal with the first question, the overall adequacy of the model. Chapter 3 deals with the contributions of each of the independent variables, and Chapter 4 focuses on testing the assumptions of the model.

In linear regression analysis, we need to know (a) whether knowing the values of all of the independent variables put together allows us to predict the dependent variable any better than if we had no information on any of the independent variables and, if so, (b) how well the independent variables, as a group, explain the dependent variable. For logistic regression, we also may be interested in the frequency of correct as opposed to incorrect predictions of the exact value of the dependent variable, in addition to how well the model minimizes errors of prediction. In linear regression, when the dependent variable is assumed to be measured on an interval or ratio scale, it would be neither alarming nor unusual to find that none of the predicted values of the dependent variable exactly matched the observed value of the dependent variable. In logistic regression, with a finite number (usually only two) of possible values of the dependent variable, we may sometimes be more concerned with whether the predictions are correct or incorrect than with how close the predicted values (the predicted conditional means, which are equal to the predicted conditional probabilities) are to the observed (0 or 1) values of the dependent variable.

R^2, F, and Sums of Squared Errors

In linear regression analysis, evaluation of the overall model is based on two sums of squares. If we were concerned with minimizing the sum of the squared errors of prediction, and if we knew only the values of the dependent variable (but not the cases to which those values belonged), we could minimize the sum of the squared errors of prediction by using the mean of Y, $\overline{Y}$, as the predicted value of Y for all cases. The sum of squared errors based on this prediction would be $\sum(Y_j - \overline{Y})^2$, the *Total Sum of Squares* or SST. If the independent variables are useful in predicting Y, then $\hat{Y}_j$, the value of Y predicted by the regression equation (the conditional mean of Y) will be a better predictor than $\overline{Y}$ of the values of Y, and the sum of squared errors $\sum(Y_j - \hat{Y}_j)^2$ will be smaller than the sum of squared errors $\sum(Y_j - \overline{Y})^2$. $\sum(Y_j - \hat{Y}_j)^2$ is called the *Error Sum of Squares* or SSE, and it is the quantity OLS selects parameters $\beta_1, \beta_2, \ldots, \beta_k$ to minimize. A third sum of squares, the *Regression Sum of Squares* or SSR, is simply the difference between SST and SSE: SSR = SST − SSE.

It is possible in a sample of cases to get an apparent reduction in error of prediction by using the regression equation instead of $\overline{Y}$ to predict the values of Y_j, even when the independent variables are really unrelated to Y. This occurs as a result of sampling variation, random fluctuations in sample values that may make it appear as though a relationship exists between two variables when there really is no relationship. The multivariate F ratio is used to test whether the improvement in prediction using $\hat{Y}$ instead of $\overline{Y}$ could be attributed to random sampling variation. Specifically, the multivariate F ratio tests two equivalent hypotheses: H_0: $R^2 = 0$ and H_0: $\beta_1 = \beta_2 = \ldots = \beta_k = 0$. For OLS linear regression, the F ratio with N cases and k independent variables can be calculated as $F = [SSR/k]/[SSE/(N - k - 1)] = (N - k - 1)SSR/(k)SSE$. The *attained statistical significance* (p) associated with the F ratio indicates the probability of obtaining an R^2 as large as the observed R^2, or β coefficients as large as the observed β coefficients, *if the null hypothesis is true.* If p is small (usually less than .05, but other values of p may be chosen), then we reject the null hypothesis and conclude that there is a relationship between the independent variables and the dependent variable that cannot be attributed to chance. If p is large, then we "fail to reject the null hypothesis" and conclude that there is insufficient evidence to be sure that the variance explained by the model is not attributable to random sample variation. This does not mean that we conclude that there is no relationship, only that, if there is a relationship, we have insufficient evidence to be confident that it exists.

The coefficient of determination, or R^2, or "explained variance" (really, the proportion of the variance that is explained), is an indicator of *substantive* significance, that is, whether the relationship is "big enough" or "strong enough" for us to be concerned about it. R^2 is a *proportional reduction in error* statistic. It measures the proportion (or, multiplied by 100, the percentage) by which use of the regression equation reduces the error of prediction, relative to predicting the mean, $\overline{Y}$. R^2 ranges from 0 (the independent variables are no help at all) to 1 (the independent variables allow us to predict the individual values Y_j perfectly). R^2 is calculated as $R^2 = \text{SSR/SST} = (\text{SST} - \text{SSE})/\text{SST} = 1 - (\text{SSE/SST})$. The F ratio and R^2 can also be expressed as functions of one another: $F = (R^2/k)/[(1 - R^2)/(N - k - 1)]$ and $R^2 = kF/(kF + N - k - 1)$.

It is possible for a relationship to be statistically significant ($p \leq .0001$) but for R^2 not to be substantively significant (e.g., $R^2 \leq .005$) for a large sample. If the independent variables explain less than $\frac{1}{2}\%$ of the variance in the dependent variable, we are unlikely to be very concerned with them, even if we are relatively confident that the explained variance cannot be attributed to random sample variation. It is also possible for a relationship to be substantively significant (e.g., $R^2 \geq .4$) but not statistically significant for a small sample. Even though the relationship appears to be moderately strong (an explained variance of .40 or, equivalently, a 40% reduction in error of prediction), there may not be enough cases for us to be confident that this result cannot be attributed to random sampling variation.

Goodness of Fit: G_M, R_L^2, and the Log-Likelihood

Close parallels to F and R^2 exist for the logistic regression model. Just as the sum of squared errors is the criterion for selecting parameters in the linear regression model, the *log-likelihood* is the criterion for selecting parameters in the logistic regression model. In presenting information on the log-likelihood, however, statistical packages usually present not the log-likelihood itself but the log-likelihood multiplied by -2. The reason for this is that the log-likelihood when multiplied by -2 has approximately a χ^2 (chi-square) distribution. For convenience and consistency with the output from computer packages such as SPSS and SAS, the log-likelihood multiplied by −2 will be abbreviated as −2LL. Because the log-likelihood is negative, the −2LL statistic is positive, and larger values indicate worse prediction of the dependent variable. The value of −2LL for the logistic regression model with only the intercept included is called the "INITIAL LOG LIKELIHOOD FUNC-TION −2 LOG LIKELIHOOD" in SPSS LOGISTIC REGRESSION, and it appears at the beginning of the logistic regression output, before the independent variables are entered into the model. In SAS, it is designated as "−2 LOG L" in the column "Intercept Only" in the output from SAS PROC LOGISTIC. The intercept-only or initial −2LL, hereafter designated D_0 to indicate that it is the −2 log-likelihood statistic with none (zero) of the independent variables in the equation, is analogous to the total sum of squares, SST, in linear regression analysis. For a dichotomous dependent variable (coded as 0 or 1), if $n_{Y=1}$ is the number of cases for which $Y = 1$, N is the total number of cases, and $P(Y = 1) = n_{Y=1}/N$ is the probability that Y is equal to 1, then

$$D_0 = -2 \left\{ n_{Y=1} \ln[P(Y = 1)] + (N - n_{Y=1}) \ln[1 - P(Y = 1)] \right\}$$

$$= -2 \left\{ (n_{Y=1}) \ln[P(Y = 1)] + (n_{Y=0}) \ln[P(Y = 0)] \right\}.$$

The value of −2LL for the logistic regression model that includes the independent variables as well as the intercept is designated as "−2 LOG LIKELIHOOD" in the "CHI-SQUARE" column in the output for SPSS LOGISTIC REGRESSION; as "−2 LOG L" in the "Intercept and Covariates" column in SAS PROC LOGISTIC; and as D, the "deviance" chi-square, by Hosmer and Lemeshow (1989). "Deviance" is a term with multiple meanings (especially when the principal example for this monograph is marijuana use), and I prefer the term "deviation" chi-square. Hereafter, this −2LL statistic will be referred to as either D_M or the deviation χ^2 for the full model. D_M is analogous to the error sum of squares, SSE, in linear regression analysis. D_M is used in logistic regression as an indicator of how *poorly* the model fits with all of the independent variables in the equation. Maintaining the analogy with linear regression, the statistical significance of D_M is analogous to the statistical significance of the *unexplained* variance in a linear regression model. It is also analogous to the goodness-of-fit χ^2 statistic in structural equation models (Jöreskog & Sörbom, 1993).

The most direct analogue in logistic regression analysis to the regression sum of squares, SSR, in linear regression is the difference between D_0 and D_M, that is, $(D_0 - D_M)$. This difference is called the "MODEL CHI-

SQUARE" in SPSS LOGISTIC REGRESSION, "-2 LOG L" in the column "Chi-Square for Covariates" in SAS PROC LOGISTIC, and G by Hosmer and Lemeshow (1989). Hereafter, it will be referred to as G_M, or the Model χ^2. G_M is analogous to the multivariate F test for linear regression, as well as the regression sum of squares. Treated as a chi-square statistic, G_M provides a test of the null hypothesis that $\beta_1 = \beta_2 = \ldots = \beta_k = 0$ for the logistic regression model. If G_M is statistically significant ($p \leq .05$), then we reject the null hypothesis and conclude that information about the independent variables allows us to make better predictions of $P(Y = h)$ (where h is some specific value, usually 1, and usually for a dichotomous dependent variable) than we could make without the independent variables. Ideally, it would be desirable to have a model in which G_M was statistically significant and D_M was not, but, especially for large samples, it is entirely possible that both G_M and D_M will be statistically significant. (For mnemonic fans, a statistically significant G_M is Good; a statistically significant D_M is Dreadful.) Given the goal of the logistic regression model (prediction of a single dependent variable) and the usefulness of the analogy between linear and logistic regression, it seems advisable to focus primarily on G_M and only secondarily on D_M.

Other indices of goodness of fit have been proposed and are available in different software packages. These include Hosmer and Lemeshow's (1989) goodness-of-fit index, $\hat{C}$, which is included in the output for SPSS LOGISTIC REGRESSION; the Score statistic; the Akaike Information Criterion (AIC); and the Schwartz Criterion (a modification of the AIC). The Score statistic, AIC, and the Schwartz criterion are provided in SAS PROC LOGISTIC. The Score statistic, discussed in Hosmer and Lemeshow (1989), is, like G_M, a test of the statistical significance of the combined effects of the independent variables in the model. The AIC and the Schwartz criterion, briefly discussed in Bollen (1989), are two related indices used for comparing models, rather than providing absolute tests of adequacy of fit. It is possible to compare the AIC or the Schwartz criterion for the fitted model with the AIC or Schwartz criterion for the model with only the intercept, but this provides little more information than G_M or D_M. Hosmer and Lemeshow's goodness-of-fit index is designed primarily as an alternative to the deviation χ^2. It is therefore of interest only to the extent that D_M is of interest. Hosmer and Lemeshow's goodness-of-fit index may have some advantage over D_M when the number of possible combinations of values of the independent variables is equal (or approximately equal) to the number of cases in the analysis but otherwise tends to produce results consistent with D_M.

Measures of Multiple Association Between
the Independent Variables and the Dependent Variable

Several analogues to the linear regression R^2 have been proposed for logistic regression. If we maintain the analogy between the $-2LL$ statistics for logistic regression and the sums of squares for linear regression analysis, the most natural choice is the analogue to SSR/SST, discussed as R_L^2 in Hosmer and Lemeshow (1989, p. 148; see also Agresti, 1990, pp. 110-111; DeMaris, 1992, p. 53; Knoke & Burke, 1980, p. 41): $R_L^2 = G_M/(D_0) = G_M/(G_M + D_M)$. Hosmer and Lemeshow present two equivalent forms of this measure, cast in terms of log-likelihoods rather than $-2LL$ statistics (Hosmer & Lemeshow, 1989, p. 148). R_L^2 is a *proportional reduction in* χ^2 or a *proportional reduction in the absolute value of the log-likelihood* measure. It indicates by how much the inclusion of the independent variables in the model reduces the badness-of-fit D_0 chi-square statistic. R_L^2 varies between 0 (for a model in which $G_M = 0$, $D_M = D_0$, and the independent variables are useless in predicting the dependent variable) and 1 (for a model in which $G_M = -2LL$ and $D_M = 0$, and the model predicts the dependent variable with perfect accuracy).

An earlier version of SAS PROC LOGISTIC, SAS (SUGI) PROC LOGIST (Harrell, 1986) included a variant of R_L^2, adjusted for the number of parameters in the model. This measure is analogous to the adjusted R^2 in linear regression, and we may denote it as R_{LA}^2 to indicate its connection with R_L^2 and to distinguish it from other R^2-type measures. $R_{LA}^2 = (G_M - 2k)/(D_0)$, where k is the number of independent variables in the model. Another alternative to R_L^2 is the pseudo-R^2 measure proposed by Aldrich and Nelson (1984) in their discussion of logit and probit models. In the notation used in this monograph, if N is the number of cases, pseudo-$R^2 = G_M/(G_M + N)$. This measure is equivalent to the squared contingency coefficient sometimes used in calculating associations for nominal variables (see Bohrnstedt & Knoke, 1988, pp. 310-311). It is equal to 0 when the independent variables are unrelated to the dependent variable (and $G_M = 0$), but a disadvantage of this measure is that it can never actually reach a value of 1, even for perfect prediction of the dependent variable. Hagle and Mitchell (1992) suggested a correction for Aldrich and Nelson's pseudo-R^2 that would allow it to vary from 0 to 1 and also noted that the corrected pseudo-R^2 provided a good approximation for the OLS regression R^2 *when the dichotomous dependent variable represented a latent interval scale*. In this instance, however, there are several other alternatives, including the possibility of using a linear probability model (because the restriction of values to a dichotomy is really artificial for a latent interval scale), using polychoric correlation and weighted least-squares estimation in the context of a more complex structural equation model (Jöreskog & Sörbom, 1993), and using R^2 itself to measure the strength of the association between the observed and predicted values of the dependent variable.

The use of R^2, the familiar coefficient of determination from ordinary least squares linear regression analysis, has received little attention in the literature on logistic regression analysis. (For an exception, see Agresti, 1990, pp. 111-112.) Its utility in logistic regression has been questioned because unlike R_L^2 and Aldrich and Nelson's pseudo-R^2, (a) it is not based on the criteria used in selecting the model parameters and (b) it can be used only when the dependent variable is dichotomous. For a dichotomous dependent variable, the mean and variance are well defined. For a polytomous (more than two categories) nominal dependent variable, the values attached to the categories are totally meaningless, and for a polytomous ordinal dependent variable with a small number of categories, ordinal measures of association are available.

There are certain advantages to the use of R^2, not instead of R_L^2 but as a supplemental measure of association between the independent variables and the dependent variable. First, using R^2 permits direct comparison of logistic regression models with linear probability, analysis of variance, and discriminant analysis models when predicting the observed value (instead of predicting the observed probability that the dependent variable is equal to that value) is of interest. Second, R^2 is relatively easy to calculate using existing statistical software. Third, R^2 is useful in calculating standardized logistic regression coefficients, a topic to be covered in the next chapter.

Assume that the dependent variable is Y and that you want to name the variable representing the value of Y predicted by the logistic regression model LPREDY. In SPSS and SAS, to obtain R^2, it is necessary to save the predicted values of the dependent variable from SPSS LOGISTIC REGRESSION [using /SAVE = PRED(LPREDY)] or from SAS PROC LOGISTIC [using OUTPUT PRED = LPREDY]. Next, use a bivariate or multiple regression routine (such as SPSS PLOT or SAS PROC REG) to calculate R^2. Alternatively, use any analysis of variance routine that calculates η^2 or η (SPSS MEANS or ANOVA; SAS PROC GLM or ANOVA), with the observed value of the independent variable, Y, as the dependent variable and the predicted value of the dependent variable, LPREDY, as the dependent variable. Because there are only two variables (the observed values of Y as one variable, the predicted values of Y as the other), $\eta^2 =$

R^2, and the two may be used interchangeably. Although for η^2 this role-switching between the dependent variable and its predicted value (which is based on the values of the independent variables) may seem strange, it exactly parallels the method for calculating canonical correlation coefficient in discriminant analysis (Klecka, 1980).

Predictive Efficiency: λ_p, τ_p, ϕ_p, and the Binomial Test

In addition to statistics regarding goodness of fit, logistic regression programs commonly print classification tables, tables that indicate the predicted and observed values of the dependent variable for the cases in the analysis. These tables resemble the contingency tables produced by SPSS CROSSTABS and SAS PROC FREQ. In most instances, we will be more interested in how well the model predicts probabilities, $P(Y_j = 1)$. In other cases, however, we may be more interested in the accurate prediction of group membership, and the classification tables may be of as much or more interest than the overall fit of the model. As of this writing, there is no consensus at all on how to measure the association between the observed and predicted classification of cases, based on logistic regression or related methods such as discriminant analysis. There are, however, several good suggestions that can easily be implemented for providing summary measures for classification tables. The best options for analyzing the prediction tables provided by logistic regression packages involve *proportional change in error* measures of the form

$$\text{Predictive efficiency} = \frac{\text{Errors without model} - \text{Errors with model}}{\text{Errors without model}},$$

$$(2.1)$$

which is a *proportional change in error* formula. If the model improves our prediction of the dependent variable, this is the same as a *proportional reduction in error* (PRE) formula. It is possible under some circumstances, however, that a model will actually do worse than chance at predicting the values of the dependent variable. When that occurs, the predictive efficiency is negative, and we have a proportional *increase* in error. The errors with the model are simply the number of cases for which the predicted value of the dependent variable is incorrect. The errors without the model differ for the three indices and depend on whether we are using a prediction, classification, or selection model.

Prediction, Classification, and Selection Models

In prediction models, the attempt is made to classify cases according to whether they satisfy some criterion such as success in college, absence of behavioral or emotional problems in the military, or involvement in illegal behavior after release from prison. There are no a priori constraints on the number or proportion of cases predicted to have or not to have the specified behavior or characteristic. In principle, it is possible (but not necessary) to have the same number of cases *predicted* to be "positive" (having the behavior or characteristic, e.g., "successes") and "negative" (not having the behavior or characteristics, e.g., "failures") as are *observed* to be positive or negative. That is, there is nothing that constrains the *marginal distributions* (the number or proportion of cases in each category, positive or negative) of predicted and observed frequencies to be equal or unequal. In particular, all cases may be predicted to belong to the same category; that is, the sample or population may be *homogeneous*. In practical terms, prediction models are appropriate when identical treatment of all groups ("lock 'em all up" or "let 'em all go") is a viable option.

In classification models, the goal is similar to that of prediction models but there is the added assumption that the cases are truly heterogeneous. Correspondingly, the evaluation of a classification model imposes the constraint that the model should classify as many cases into each category as are actually observed in each category. The proportion or number of cases observed to be in each category (the *base rate*) should be the same as the proportion or number of cases predicted to be in each category. To the extent that a model fails to meet this criterion, it fails as a classification model. Complete homogeneity is an unacceptable solution for a classification model. Practically speaking, classification models are appropriate when heterogeneity is assumed and identical treatment of all groups is not a viable option.

In selection models (Wiggins, 1973), the concern is with "accepting" or "rejecting" cases for inclusion in a group, based both on whether they will satisfy some criterion for success in the group and on the minimum required, maximum allowable, or specified number of cases that may (or must) be included in the group. The proportion of cases observed to be successful (the *base rate* again) may or may not be equal to the proportion of cases accepted or selected for inclusion in the group (the *selection ratio*). For example, a company may need to fill 20 positions from a pool of 200 applicants. The selection ratio will be 20/200 = .10 (10%) regardless of whether the base rate (the observed probability of success on the job) is

5% or 20%, half or twice the selection ratio. The classification tables provided in logistic regression packages may naturally be regarded as prediction or classification models. They may be used to construct selection models, but they must be altered (unless, purely by coincidence, the selection ratio turns out to be equal to the base rate) so that the correct number of cases is selected.

Common Measures of Association for Contingency Tables as Indices of Predictive Efficiency

Among the various measures that have been considered as indices of predictive efficiency are several measures of association that are commonly employed to analyze contingency tables: phi, Goodman and Kruskal's gamma, kappa, the contingency coefficient, Pearson's r, and the odds ratio (Farrington & Loeber, 1989; Mieczkowski, 1990; Ohlin & Duncan, 1949). The problem with using common contingency table measures of association to analyze 2×2 or larger prediction tables lies in the distinction between (a) the strength of a relationship between an independent variable X and a dependent variable Y and (b) the strength of the relationship between predicted group membership $E(Y_j)$ and observed group membership Y_j. These differences are illustrated in Figure 2.1. Table A in Figure 2.1 represents the general format to be used throughout this section in designating cell and marginal frequencies in 2×2 tables, Table B represents the hypothetical relationship between ethnicity and political orientation, and Table C illustrates the hypothetical relationship between predicted and observed political orientation.

Although Tables B and C are numerically identical, the inferences to be drawn from them are very different. In Table B, knowledge of ethnicity allows us to predict political orientation with a *proportional reduction in error* or PRE (Bohrnstedt & Knoke, 1988; Costner, 1965) of .20 according to Goodman and Kruskal's lambda or .04 according to Goodman and Kruskal's tau. In Table C, the PRE is the same—but only if we predict the *opposite* of what the hypothetical model predicts. Actually, the model does *worse* than chance in predicting political orientation (a situation that may arise naturally with skewed data, or with the application of a prediction model developed from one set of data to another set of data). If every case were *misclassified*, both lambda and tau would have a value of 1.00 for Table C; they would make no distinction between perfectly accurate classification and perfect *misclassification*. Pearson's r and its equivalents for 2×2 tables, Kendall's tau and phi when it is calculated as

Table A: Standard Format for Prediction Tables

Predicted Y

		Positive (success)	Negative (failure)	
Observed Y	Positive (success)	a	b	$a + b$
	Negative (failure)	c	d	$c + d$
		$a + c$	$b + d$	$a + b + c + d$

Table B: Ethnicity and Political Orientation

X: Ethnicity

		European	Non-European	
Political Orientation	Conservative	20	30	50
	Liberal	30	20	50
		50	50	100

Table C: Predicted and Observed Political Orientation

Predicted Political Orientation

		Conservative	Liberal	
Observed Political Orientation	Conservative	20	30	50
	Liberal	30	20	50
		50	50	100

Figure 2.1. Association Versus Prediction

For Tables B and C, Goodman and Kruskal's $\lambda = .20$; Goodman and Kruskal's $\tau = \phi^2 = r^2 = .04$.

$$\phi = \frac{ad - bc}{\sqrt{(a + b)(a + c)(b + d)(c + d)}},$$

would indicate misclassification with a negative sign and may be interpreted as PRE measures when squared. For larger tables with unordered categories, however, Pearson's r and Kendall's tau cannot be used, and phi becomes Cramer's V, which no longer has a PRE interpretation. The odds ratio may also be used for 2×2 tables but, for larger tables, two or more odds ratios must be calculated, and the odds ratio no longer provides a single summary measure of accuracy of prediction. On the whole, it does not appear that the application of common measures of association for contingency tables to predictive tables provides a straightforward or general solution to the problem of estimating accuracy of prediction. Pearson's r and r^2, or ϕ and ϕ^2, are reasonable indices for use with 2×2, but not larger, tables, as long as we remember to interpret them contingent upon the sign of r or ϕ.

λ_p, τ_p, and ϕ_p

Equation 2.1 provides a basic form for indices of predictive efficiency. Errors with the model is simply the number of cases misclassified when we use the model and is analogous to the error sum of squares. Errors without the model is analogous to the total sum of squares and depends on whether we are using a prediction model, a classification model, or a selection model. For a prediction model, the approach most closely analogous to linear regression (with an interval-level dependent variable) is to use the mode of the dependent variable as the predicted value for all cases (analogous to using the mean in linear regression). This method of defining errors without the model is the same as the one used in defining Goodman and Kruskal's λ for contingency tables with nominal variables, and it gives us an index first proposed by Ohlin and Duncan (1949). Because of the similarity to Goodman and Kruskal's λ, the index is here referred to as λ_p or lambda-p, where the subscript p refers to its use with prediction tables.

Lambda-p is a proportional reduction in error (PRE) measure like R^2 when it is positive but, if the model does worse than predicting the mode, λ_p may be negative, indicating the proportional *increase* in error. The possible values of λ_p vary depending on the marginal distributions. In general, the full range of possible values for λ_p in all tables with N cases is from $1 - N$ to 1.

For a classification model, an appropriate definition of the expected error without the model is

$$\text{Errors without model} = \sum_{i=1}^{N} f_i \left[(N - f_i)/N \right]$$

where N is the sample size and f_i is the number of cases observed in category i. This is the same formula for error without the model as is used for Goodman and Kruskal's τ. An index based on this definition of errors without the model was proposed by Klecka (1980) for use with discriminant analysis models. Parallel to λ_p, Klecka's index will be referred to as tau-p (τ_p), or tau for prediction tables.

Like λ_p, τ_p is a measure of change. Unlike λ_p, τ_p requires that, even in the estimation of error without the model, cases must be separated into distinct groups or categories, and not all placed in the same category. In effect, τ_p adjusts the expected number of errors for the base rates of classification. Accuracy of prediction is thus secondary, subject to the a priori assumption of heterogeneity. As with λ_p, a value of 1 for τ_p indicates that all cases are correctly classified, and a negative value for τ_p indicates that the prediction model does worse than expected (based on the observed marginal distribution) in predicting the classification of cases. Secondary properties of τ_p include the fact that $\tau_p \geq \lambda_p$ because the number of errors without the model for τ_p will be equal to or larger than the number of errors without the model for λ_p. For tables with equal marginal distributions, τ_p varies between -1 and $+1$, but the maximum value of τ_p is less than 1 when the marginal distributions are unequal. In the worst possible case, with extremely skewed and inconsistent marginal distributions, the minimum value of τ_p is equal to $1 - [N^2/(2N - 2)]$. The range of τ_p is not constant but varies from 1 to 2 for different marginal distributions. A smaller range occurs for models in which τ_p is negative and large in absolute value.

It is also possible to construct a proportional change in error measure of accuracy of prediction for selection models. For such a measure, the error with the model will be $b + c$, just as it is for λ_p and τ_p. Error with the model should depend on both the base rate, $B = (a + b)/N$, and the selection ratio, $S = (a + c)/N$. Given B, S, and N, we know the expected value of cell a (Table A, Figure 2.1): $E(a) = BSN$. Because a 2×2 prediction table has only 1 degree of freedom, once the expected value of a is known then, given the marginal distribution, the expected values of all of the other cells are known and are identical to the expected values used in calculating the χ^2 statistic. The expected error is $E(b + c) = [(a + b)(b + d)/N + (c + d)(a + c)/N]$. Plugging these values into the PRE formula, we obtain a proportional change in error measure, ϕ_p:[8]

$$\phi_p = \frac{[(a+b)(b+d)/N] + [(c+d)(a+c)/N] - (b+c)}{[(a+b)(b+d)/N] + [(c+d)(a+c)/N]}$$

$$= \frac{(a+b)(b+d)+(c+d)(a+c)-N(b+c)}{(a+b)(b+d)+(c+d)(a+c)}$$

$$= \frac{(ad-bc)}{\frac{1}{2}[(a+b)(b+d)+(c+d)(a+c)]}.$$

For tables with equal marginal distributions, ϕ_p has a maximum value of +1. In general, it varies between −1 and +1, but the actual maximum, minimum, and range depend on the marginal distributions. As long as errors without the model are calculated as the sum of the expected frequencies in cells b and c (Table A, Figure 2.1) and errors with the model are calculated as the sum of the observed frequencies in cells b and c, ϕ_p can be extended to tables larger than 2×2 and still retain a proportional change in error interpretation. For 2×2 tables, it can be shown that $|\phi_p| \leq |\phi|$, and that ϕ_p has the same sign as ϕ and Pearson's r (the numerator is the same, $ad - bc$, as for ϕ). When all cases are correctly predicted, $\phi_p = 1$. Otherwise, $\phi_p < 1$, even when the maximum possible number of cases for a given set of marginals is correctly classified.[9]

Statistical Significance of λ_p, τ_p, and ϕ_p

Lambda-p, tau-p, and phi-p are analogous to R^2 as measures of substantive significance. For statistical significance, an analogue to the F test is the normal approximation to the binomial test. Let N = total sample size, P_e = (errors without model)/N, and p_e = (errors with model)/N. The binomial statistic d may then be computed as

$$d = \frac{(P_e - p_e)}{\sqrt{P_e(1 - P_e)/N}}$$

and d is approximately normally distributed (Bulmer, 1979).[10] Note that what is being compared is not the *proportion of cases in each category* but the *proportion of cases correctly or incorrectly classified* by the model. This test is the same for λ_p, τ_p, and ϕ_p, for predictive, classification, and selection models. Only the definition of errors without the model differs.

In the proposed test of statistical significance, the value of the observed classification is taken as given; the test indicates whether the proportion incorrectly predicted with the model (which is, by assumption, dependent on the model, and thus variable) differs significantly from the proportion incorrectly predicted without the model (which is dependent only on the marginal distribution, not on the model, and thus assumed to be fixed). This form of the binomial test, which explicitly uses the expected number of errors as the criterion by which the number of errors generated by the model is to be judged, is preferable to the binomial test for a *difference* of two proportions (Bulmer, 1979, p. 145), which assumes that the two proportions (errors with the model and errors without the model) are based on separate samples (possibly of unequal size). This separate-samples condition is clearly not met when comparing observed and predicted classifications taken from the classification tables generated by logistic regression, or expected and actual errors, both of which are derived from these tables. The binomial test for the difference of two proportions may, however, be useful if we want to test whether the overall predictive accuracy (percentage correctly predicted) is statistically significantly different for two separate prediction models. Even in this situation, however, we would want a separate test to indicate whether either or both of the prediction models was significantly better than chance in reproducing the observed classification of cases.

Other Proposed Indices of Predictive Efficiency

Maddala (1983, pp. 76-77) reviewed three indices of predictive efficiency proposed in the econometric literature. One he dismissed (appropriately) as being unable to distinguish a perfectly accurate model from a perfectly inaccurate model, a characteristic it shares with many of the measures of association commonly used in the analysis of contingency tables. A second, which considered both "first best" and "second best" guesses, was more akin to indices of goodness of fit, insofar as a "near miss" (the second most likely category, according to the prediction model) was credited as an accurate choice. The third typically varied from -1 to $+.50$, depending on the marginal distribution, and produced values similar to ϕ and Pearson's r. Because it lacks a PRE interpretation, it appears to have little or no advantage over ϕ or Pearson's r.

Copas and Loeber (1990), Farrington and Loeber (1989), and Loeber and Dishion (1983) proposed a measure they called Relative Improvement Over Chance, or RIOC. Although Loeber and his colleagues applied the RIOC to the analysis of prediction and classification tables, the measure corrects for differences between the base rate and the selection ratio, an approach appropriate for selection models, rather than classification or prediction models. This measure is identical to the coefficient ϕ', the ϕ

coefficient corrected for the marginal distribution. Unlike ϕ, ϕ' has no PRE interpretation. The measure varies between -1 (for perfectly inaccurate prediction, *if cells b and c in Table A of Figure 2.1 are both nonzero*) and $+1$. If *either one* of the cells (*b* or *c*) containing incorrect predictions is equal to zero, RIOC $= 1$, *regardless of how small a proportion of the cases are correctly classified* (a problem it shares with Yule's Q, a measure of association sometimes used for contingency tables). Even if over 90% of the cases are misclassified, RIOC may have a value of 1.[11]

Examples: Assessing the
Adequacy of Logistic Regression Models

One reason for the lack of consensus about indices of predictive efficiency may be the fact that researchers are more often interested in the goodness of fit of the model (as indicated by G_M and R_L^2) than in the accuracy of prediction or classification of the model, as indicated by the classification table and indices such as λ_p, τ_p, and ϕ_p. Especially for theory testing, goodness of fit is simply more important than accuracy of classification. The amount of space devoted to accuracy of prediction in this monograph reflects the relative lack of development in this area, rather than its importance, compared to the assessment of goodness of fit. Often the two approaches, goodness of fit and accuracy of prediction, will produce consistent results. It is entirely possible, however, to have a model that fits well but does a poor job of predicting category membership.

Figures 2.2 and 2.3 illustrate how indices of goodness of fit and predictive efficiency may lead to very different substantive conclusions. In Figure 2.2, hypothetical data are presented for a single dependent variable, TRUE, and a single predictor, P1. The standard output has been edited to include R_L^2, R^2, λ_p, and τ_p. For the 40 cases analyzed in Figure 2.2, the model fits well. $G_M =$ Model $\chi^2 = 20.123$ and is statistically significant (significance $= p = .0000$), leading us to reject the null hypothesis that the independent variable, P1, is not related to the dependent variable, TRUE. $R_L^2 = .363$, suggesting a moderate association between TRUE and P1. The binomial d is the same for both lambda-p and tau-p (50% expected error for both): $d = 5.060$, with statistical significance $p = .000$. Both tau-p and lambda-p are equal to .80, indicating that the independent variable allows us to classify the cases (into the categories of the dependent variable) with a very high degree of accuracy, as reflected in the classification table. Overall, the accuracy of prediction is considerably higher than the ability of the model to predict the probability $P(Y_j = 1)$. The plot of observed groups and

33

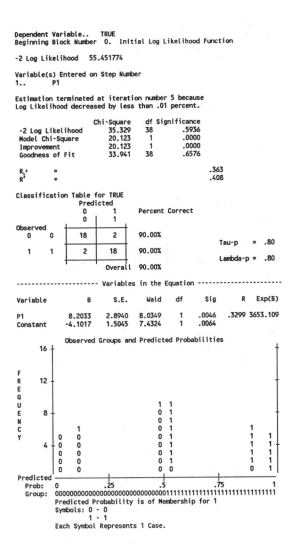

```
Dependent Variable..    TRUE
Beginning Block Number  0.  Initial Log Likelihood Function

-2 Log Likelihood    55.451774

Variable(s) Entered on Step Number
1..    P1

Estimation terminated at iteration number 5 because
Log Likelihood decreased by less than .01 percent.

                     Chi-Square   df  Significance
-2 Log Likelihood      35.329     38     .5936
Model Chi-Square       20.123      1     .0000
Improvement            20.123      1     .0000
Goodness of Fit        33.941     38     .6576

R₁²      =                               .363
R₂²      =                               .408

Classification Table for TRUE
                    Predicted
                    0    1     Percent Correct
                    0    1
Observed
    0   0  │  18  │   2  │     90.00%
                                            Tau-p    = .80
    1   1  │   2  │  18  │     90.00%
                                            Lambda-p = .80
            Overall        90.00%

--------------------- Variables in the Equation ----------------------

Variable      B        S.E.    Wald     df    Sig     R      Exp(B)

P1         8.2033    2.8940   8.0349    1    .0046   .3299  3653.109
Constant  -4.1017    1.5045   7.4324    1    .0064
```

Observed Groups and Predicted Probabilities

Figure 2.2. Logistic Regression Output for Hypothetical "Good Prediction" Data

predicted probabilities at the bottom of Figure 2.2 indicates that the predicted probabilities are sometimes very high and sometimes close to .5, the cutoff for classification into $(Y = 1)$ or $(Y = 0)$. Accuracy of prediction is very high, even for the cases whose predicted probability of belonging in $(Y = 1)$ is close to .5.

34

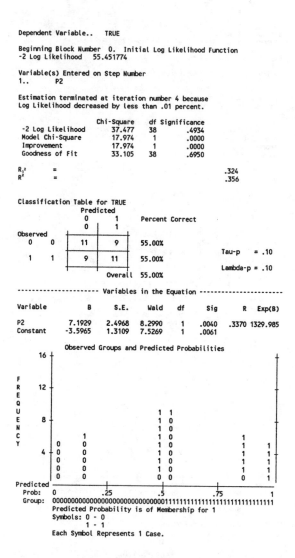

```
Dependent Variable..   TRUE

Beginning Block Number  0.  Initial Log Likelihood Function
-2 Log Likelihood   55.451774

Variable(s) Entered on Step Number
1..    P2

Estimation terminated at iteration number 4 because
Log Likelihood decreased by less than .01 percent.

                    Chi-Square   df  Significance
-2 Log Likelihood     37.477     38     .4934
Model Chi-Square      17.974      1     .0000
Improvement           17.974      1     .0000
Goodness of Fit       33.105     38     .6950

R₁²    =                                          .324
R²     =                                          .356


Classification Table for TRUE
                Predicted
                0     1      Percent Correct
                0     1
Observed
  0   0       11     9       55.00%
                                              Tau-p    = .10
  1   1        9    11       55.00%
                                              Lambda-p = .10
              Overall        55.00%

---------------------- Variables in the Equation ----------------------

Variable         B       S.E.     Wald    df    Sig      R    Exp(B)

P2            7.1929    2.4968   8.2990    1   .0040  .3370 1329.985
Constant     -3.5965    1.3109   7.5269    1   .0061

          Observed Groups and Predicted Probabilities
     16 +                                                     +

 F
 R   12 +                                                     +
 E
 Q
 U                                  1  1
 E    8 +                           1  0                      +
 N                                  1  0
 C           1                      1  0               1
 Y           0    0                 1  0               1    1
      4 +    0    0                 1  0               1    1 +
             0    0                 1  0               1    1
             0    0                 1  0               1    1
             0    0                 0  0               0    1
Predicted  ----+------------+------------+------------+----
   Prob:   0           .25          .5          .75          1
   Group:  000000000000000000000000000111111111111111111111111111111
           Predicted Probability is of Membership for 1
           Symbols: 0 - 0
                    1 - 1
           Each Symbol Represents 1 Case.
```

Figure 2.3. Logistic Regression Output for Hypothetical "Poor Prediction" Data

In Figure 2.3, G_M is again statistically significant, and both R_L^2 and R^2 indicate a moderately strong relationship between the dependent variable, TRUE, and the new predictor, P2. However, the binomial $d = .632$, with statistical significance $p = .264$ (one-tailed), suggesting that the classification on the dependent variable is not related to the values of the independent variable. Tau-p and lambda-p indicate no more than a weak relationship between the observed and predicted classification of the cases. The plot of observed groups and predicted probabilities indicates why. Now, instead of accurate prediction when the predicted probability is close to .5, predictions close to .5 are nearly all inaccurate. For 26 of the 40 cases, the predicted probabilities are the same for P1 and P2. For the other 14, the predicted probability changed by .02, either from .49 to .51 or from .51 to .49. This had little impact on the overall goodness of fit of the model, as measured by R_L^2, R^2 or η^2, and G_M, but it had a tremendous impact on the indices of predictive efficiency.

Figure 1.4, first discussed in Chapter 1, presented the results of a bivariate logistic regression analysis for real data, the relationship between exposure to delinquent friends (EDF5) and prevalence of marijuana use (PMRJ5). From Figure 1.4, we can reject the null hypothesis that EDF5 is unrelated to PMRJ5, based on G_M (significance = .0000). Notice that D_M, the deviation χ^2 (−2 log-likelihood) is not statistically significant, and neither is the Hosmer and Lemeshow "Goodness of Fit" measure, indicating that the model with only EDF5 as a predictor fits the data well. The prediction table appears to indicate fairly good accuracy of prediction, but we need to calculate lambda-p and/or tau-p to get a quantitative estimate of how well the cases are classified by the model. Finally, notice that the predicted values from the model have been saved as a new variable, LPEPMRJ5 (*Logistic regression Prediction from EDF5 of PMRJ5*). This permits us to use a separate analysis of variance or bivariate regression routine to calculate R^2.

From the data in Figure 1.4, we can directly calculate R_L^2 = (Model Chi-Square)/(Initial Log-Likelihood Function −2 Log-Likelihood) = $(213.947)/(299.30563) = .715$. Lambda-p is equal to the number of cases in the smaller observed category ($Y = 1$: $37 + 44 = 81$) minus the number of cases incorrectly predicted by the model ($37 + 14 = 51$), divided by the number of cases in the smaller category, so lambda-p = $(81 − 51)/81 = .370$. This is a moderately strong reduction in the error of *prediction*. Tau-p is a bit more complicated to calculate. First find the sum of the cases in each category for the observed value of Y: for $Y = 0$, $n_{Y=0} = 150$, and for $Y = 1$, $n_{Y=1} = 81$. For a dichotomous dependent variable, the expected number of errors is the product of the two sums, divided by the total number of cases (231), and multiplied by two (because we expect the same number of errors in each category for a dichotomous variable): $(2)(150)(81)/231 = 103.9$.

Tau-p is the expected number of errors minus the actual number of errors (51), divided by the expected number of errors: tau-p = (103.9 − 51)/103.9 = .509. This indicates that the model reduces the error of *classification* of cases as users or nonusers of marijuana by over half.

We can use the expected errors for lambda-p and tau-p to calculate the binomial d statistic for each measure. For lambda-p, the expected number of errors is 81, corresponding to a proportion of 81/231 = .351, and the observed number of errors is 51, corresponding to a proportion of 51/231 = .221; therefore $d = (P_e - p_e)/\sqrt{P_e(1 - P_e)/N} = (.351 - .221)/\sqrt{(.351)(.649)/231} = 4.140$, with statistical significance $p = .000$. For tau-p, without going into as much detail, $d = 6.996$, with statistical significance $p = .000$. Finally, to calculate R^2 or η^2, we must use the results of an analysis of variance or a bivariate regression routine.

We can directly compare the results of the logistic regression analysis in Figure 1.4 with the results of the linear regression analysis, with the same variables, in Part C of Figure 1.1. In particular, the explained variance for PMRJ5 in the logistic regression model, $R^2 = .34$, is actually slightly higher than the explained variance for PMRJ5 in the linear regression model, $R^2 = .32$ (from Part C of Figure 1.1).[12] This occurs despite the fact that the linear regression model tries to maximize R^2 (by minimizing the sum of the squared errors), and the logistic regression model does not. It appears that the logistic regression model fits the data well, indicates a moderately strong relationship between the predictor and the dependent variable, and does a fairly good job of predicting the classification of the cases.

Conclusion: Summary Measures for Evaluating the Logistic Regression Model

In linear regression, we use the F statistic and R^2 to test statistical significance and substantive significance, respectively, of the relationship between the dependent variable and the independent variables. Both are based on the total and error sums of squares, SST and SSE. In logistic regression, if our principal concern is with how well the model fits the data (e.g., in the context of theory testing), we use G_M and R_L^2, based on −2LL, to test for statistical and substantive significance. If our concern is less with the overall fit of the model and more with the accuracy with which the model predicts actual category membership on the dependent variable, the binomial d and one of the three indices of predictive efficiency (λ_p, τ_p, or ϕ_p) are used to assess the statistical and substantive significance of the model. For most purposes, τ_p is likely to be most appropriate, but careful attention to the purpose of the model (prediction, classification, or selection) is necessary to choose among the three indices. These indices may also be applied to prediction tables generated by procedures other than logistic regression and are not limited to dichotomous variables; they are applicable to any prediction table in which correct predictions can be distinguished from incorrect predictions.

With the exception of G_M, none of the aforementioned indices is routinely available (as this is being written) in such widely used logistic regression software as SPSS LOGISTIC REGRESSION or SAS PROC LOGISTIC. It is relatively easy, however, to compute them by hand, or to save predicted values for use in other statistical routines. For this and for other purposes (see Chapter 4), it is sometimes necessary to work around the limitations of presently existing software for logistic regression analysis.

3. INTERPRETING THE
LOGISTIC REGRESSION COEFFICIENTS

In linear regression analysis, we evaluate the contribution of each independent variable to the model by testing for its statistical significance and then examining the substantive significance of its effect on the dependent variable. Statistical significance is evaluated using an F or t statistic to produce a probability (p) that we would find this strong a relationship in a sample this large if there really were no relationship between the independent variable and the dependent variable. Substantive significance may be evaluated in one of several ways. We may examine the unstandardized regression coefficient to see whether the change in the dependent variable associated with a given amount of change in the independent variable is large enough to be concerned about. (The words "associated with" are used here in preference to language that would imply a causal relationship, because the relationships described here are definitely predictive and may, but need not be, causal in nature.) In order to apply this test, we must have some idea beforehand how big a change needs to be before we are concerned with it and, equivalently, how big a change we are willing to ignore. Unstandardized regression coefficients are especially useful for evaluating the practical impact of one variable on another and for comparing the effects of the same variable in different samples.

Alternatively, especially when there are no clear criteria for deciding "how big is big," and when some variables are not measured in natural units of measurement (feet, pounds, dollars) but are instead scale scores (an example of this would be the variable EDF5, exposure to delinquent friends), we may focus on a *standardized* regression coefficient, which indicates how many *standard deviations* a dependent variable changes in response to a 1 standard deviation change in the independent variable. Use of standardized coefficients is especially appropriate for theory testing and when the focus is on comparing the effects of different variables for the same sample.

For some purposes, stepwise methods are used to evaluate the contribution of variables to the regression equation. This is especially the case when we test for nonlinearity (e.g., by including quadratic terms) or nonadditivity (by including interaction terms) in the regression equation. The decision about whether the inclusion of nonlinear or nonadditive terms is justified is typically based on the magnitude and statistical significance of the change in the explained variance, R^2. Stepwise methods are also used in exploratory analysis, when we are more concerned with theory development than theory testing. Such research may occur in the early stages of the study of a phenomenon, when neither theory nor knowledge about correlates of the phenomenon is well developed. Criteria for stepwise inclusion or removal of variables for a model generally involve tests that are similar to but less restrictive than the tests used in theory testing.

Statistical Significance in Logistic Regression Analysis

Several methods have been used to evaluate the statistical significance of the contribution of an independent variable to the explanation of a dependent variable. One stands out as being clearly the best, in the sense of being the most accurate: the likelihood ratio test. In the likelihood ratio test, the logistic regression model is calculated with and without the variable being tested. The likelihood ratio test statistic is equal to G_M for the model with the variable minus G_M for the model without the variable. The result, which we can call G_1 when we test X_1, G_2 when we test X_2, . . . , G_k when we test X_k, has a chi-square distribution with degrees of freedom equal to the degrees of freedom in the model with X minus the degrees of freedom in the model without X. For example, if we designate G_{M1} to represent the model chi-square with X_k in the model, and G_{M2} to represent the model chi-square with X_k not in the model, $G_k = G_{M1} - G_{M2}$, and, if X is a continuous, interval, or ratio variable, then G_k has 1 degree of freedom.

The only drawback to the use of the likelihood ratio statistic is that it requires more time to compute than alternative tests for statistical significance. If you are paying for every second on a mainframe computer, this may be a serious concern, but for many users with access to relatively fast personal computers and workstations, this is irrelevant except for very large samples. Nonetheless, statistical packages are often written to use a less computationally intensive alternative to the likelihood test, the Wald statistic, to test for the statistical significance of individual coefficients. In Figure 1.4, the Wald statistic appears following the coefficient (B) and its standard error (S.E.). The Wald statistic may be calculated as $W_k^2 = [b_k/$ (standard error of $b_k)]^2$, in which case it is asymptotically distributed as a chi-square distribution, or as $W_k = b_k/$(standard error of b_k), in which case it follows a standard normal distribution (Hosmer & Lemeshow, 1989, p. 31; SAS Institute, Inc., 1989, p. 1097; SPSS, Inc., 1991, pp. 140-141) and its formula parallels the formula for the t ratio for coefficients in linear regression. The disadvantage of the Wald statistic is that, for large b, the estimated standard error is inflated, resulting in failure to reject the null hypothesis when the null hypothesis is false.[13]

Figure 3.1 presents SPSS output for a dependent variable with four predictors. PMRJ5, the prevalence of marijuana use, is again the dependent variable, and EDF5, exposure to delinquent friends, is again included as a predictor. BELIEF4 is a scale measuring how wrong (very wrong, wrong, a little wrong, not wrong at all) the respondent believes it is to commit each of several illegal acts (assault, theft, selling hard drugs, etc.), parallel to the items used in constructing EDF5. BELIEF4 is measured immediately prior to the period for which data were collected on prevalence of marijuana use. SEX is coded 0 for females and 1 for males.

The coding for ETHN is given at the beginning of Figure 3.1. The first coefficient for ETHN corresponds to being African American ("Black") and the second to being other than non-Hispanic European American or African American ("Other"). The number in the column (1) or (2) corresponds to the number of the coefficient that is set to 1 for individuals falling into each row. Thus the first coefficient for ETHN is multiplied by 1 for African Americans and by 0 otherwise and the second coefficient is multiplied by 1 for Other ethnic groups and by 0 otherwise. This is an example of the use of a set of "dummy" or design variables in logistic regression to represent a single categorical variable and is parallel to the use of dummy variables or design variables in linear regression (Hardy, 1993; Lewis-Beck, 1980).

In the section of Figure 3.1 labeled "Variables in the Equation," we find logistic regression coefficients, standard errors, Wald statistics (W_k^2), the degrees of freedom (df) associated with each variable, and the statistical significance of the Wald statistic. From Figure 3.1, it appears that EDF5, BELIEF4, and SEX have statistically significant effects on PMRJ5. For

```
Total number of cases:        257 (Unweighted)
Number rejected because of missing data:  30
Number of cases included in the analysis: 227
```

Dependent Variable Encoding:

```
Original      Internal
Value         Value
   .00        0
  1.00        1
```

```
                              Parameter
                 Value  Freq  Coding
                                (1)    (2)
ETHN
  White            1    175   .000   .000
  Black            2     37  1.000   .000
  Other            3     15   .000  1.000
SEX
  Female           0    117   .000
  Male             1    110  1.000
```

Dependent Variable..   PMRJ5

Beginning Block Number 0. Initial Log Likelihood Function -2 Log Likelihood   294.61587

* Constant is included in the model.

Beginning Block Number 1. Method: Enter

Variable(s) Entered on Step Number 1..   EDF5 BELIEF4 SEX ETHN

Estimation terminated at iteration number 4 because Log Likelihood decreased by less than .01 percent.

```
                    Chi-Square   df Significance
-2 Log Likelihood     186.359   221    .9564
Model Chi-Square      108.257     5    .0000
Improvement           108.257     5    .0000
Goodness of Fit       298.335   221    .0004
```

Classification Table for PMRJ5

```
                   Predicted
                  no      yes    Percent Correct
                  n   |    y
              +-------+-------+
Observed
   no      n   |  134  |   13  |   91.16%
              +-------+-------+
   yes     y   |   28  |   52  |   65.00%
              +-------+-------+
                       Overall  81.94%
```

---------------------- Variables in the Equation ----------------------

Variable	B	S.E.	Wald	df	Sig	R	Exp(B)
EDF5	.4067	.0694	34.3414	1	.0000	.3313	1.5019
BELIEF4	-.1178	.0596	3.9026	1	.0482	-.0804	.8888
SEX(1)	-1.5143	.4046	14.0077	1	.0002	-.2019	.2200
ETHN			1.1897	2	.5516	.0000	
ETHN(1)	.2449	.5079	.2324	1	.6297	.0000	1.2775
ETHN(2)	.7716	.7446	1.0739	1	.3001	.0000	2.1633
Constant	-1.7493	2.0283	.7438	1	.3884		

Figure 3.1. SPSS LOGISTIC REGRESSION Output

ETHN, the Wald statistic is computed for the variable as a whole and also separately for each of the coefficients corresponding to the separate categories of ethnicity. The effect of ETHN on PMRJ5 does not appear to be statistically significant, and neither does the intercept (Constant). Toward

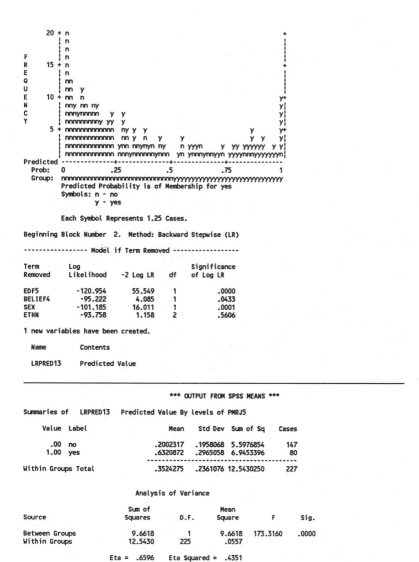

```
      20 + n                                                                    +
         | n                                                                    |
         | n                                                                    |
 F       | n                                                                    |
 R       15 + n                                                                  +
 E       | n                                                                    |
 Q       | mn                                                                   |
 U       | mn   y                                                               |
 E      10 + mn   n                                                            y+
 N       | nny nn ny                                                          y|
 C       | nnnynnnn   y   y                                                   y|
 Y       | nnnnnnnny yy   y                                                   y|
         5 + nnnnnnnnnnn  ny y  y                                     y        y+
         | nnnnnnnnnnnn  nn y  n    y      y              y  y       y|
         | nnnnnnnnnnnn  ynn nnymyn ny    n yyyn   y  yy yyyyyy  y  y|
         | nnnnnnnnnnnn  nnnynnnnnynnn   yn ynnnynnyyn yyyynnnyyyyyyyn|
Predicted ----------------+----------------+----------------+----------------
Prob:    0         .25          .5         .75          1
Group:   nnnnnnnnnnnnnnnnnnnnnnnnnnnnnnnnnnnnnyyyyyyyyyyyyyyyyyyyyyyyyyyyyyyyy
         Predicted Probability is of Membership for yes
         Symbols: n - no
                  y - yes
```

Each Symbol Represents 1.25 Cases.

Beginning Block Number 2. Method: Backward Stepwise (LR)

```
---------------- Model if Term Removed ------------------
```

Term Removed	Log Likelihood	-2 Log LR	df	Significance of Log LR
EDF5	-120.954	55.549	1	.0000
BELIEF4	-95.222	4.085	1	.0433
SEX	-101.185	16.011	1	.0001
ETHN	-93.758	1.158	2	.5606

1 new variables have been created.

Name	Contents
LRPRED13	Predicted Value

*** OUTPUT FROM SPSS MEANS ***

Summaries of   LRPRED13   Predicted Value By levels of PMRJ5

Value	Label	Mean	Std Dev	Sum of Sq	Cases
.00	no	.2002317	.1958068	5.5976854	147
1.00	yes	.6320872	.2965058	6.9453396	80
Within Groups Total		.3524275	.2361076	12.5430250	227

Analysis of Variance

Source	Sum of Squares	D.F.	Mean Square	F	Sig.
Between Groups	9.6618	1	9.6618	173.3160	.0000
Within Groups	12.5430	225	.0557		

Eta = .6596   Eta Squared = .4351

Figure 3.1. Continued

the end of Figure 3.1, just before the output from SPSS MEANS, under the
heading "Model if Term Removed," likelihood ratio statistics were ob-

The LOGISTIC Procedure

Data Set: WORK.DATA1
Response Variable: PMRJ5
Response Levels: 2
Number of Observations: 227
Link Function: Logit

Response Profile

Ordered Value	PMRJ5	Count
1	1	80
2	0	147

WARNING: 30 observation(s) were deleted due to missing values for the response or explanatory variables.

Criteria for Assessing Model Fit

Criterion	Intercept Only	Intercept and Covariates	Chi-Square for Covariates
AIC	296.616	198.359	.
SC	300.041	218.909	.
-2 LOG L	294.616	186.359	108.257 with 5 DF (p=0.0001)
Score	.	.	89.457 with 5 DF (p=0.0001)

Analysis of Maximum Likelihood Estimates

Variable	DF	Parameter Estimate	Standard Error	Wald Chi-Square	Pr > Chi-Square	Standardized Estimate	Odds Ratio
INTERCPT	1	-1.7498	2.0285	0.7441	0.3883	.	0.174
EDF5	1	0.4068	0.0694	34.3468	0.0001	0.954476	1.502
BELIEF4	1	-0.1179	0.0597	3.9033	0.0482	-0.256713	0.889
SEX	1	-1.5148	0.4047	14.0130	0.0002	-0.418313	0.220
BLACK	1	0.2451	0.5080	0.2327	0.6295	0.050013	1.278
OTHER	1	0.7720	0.7446	1.0748	0.2999	0.105968	2.164

Association of Predicted Probabilities and Observed Responses

Concordant = 88.2%	Somers' D = 0.766
Discordant = 11.6%	Gamma = 0.767
Tied = 0.2%	Tau-a = 0.351
(11760 pairs)	c = 0.883

Figure 3.2. SAS PROC LOGISTIC Output

tained for the variables in the model.[14] Substantively, the conclusions remain the same, and the significance levels are very similar for the Wald and likelihood ratio statistics.

Figure 3.2 provides partial output from SAS for the same model that was estimated in Figure 3.1. One difference between the SAS output in Figure 3.2 and the SPSS output in Figure 3.1 is the treatment of the categorical variable ETHN, which in SAS must be separated into design variables before it is entered into the analysis, because SAS assumes that the independent variables in PROC LOGISTIC have true numeric values (SAS

Institute, Inc., 1989, p. 1079). Also, there is no test for ETHN as a single variable; each of the design variables is tested separately. Such a test could be performed by analyzing separate models, one with and one without ETHN as a predictor. Otherwise, however, the parameter estimates, standard errors, Wald statistics, p values, and -2 log-likelihood statistics are practically identical to those produced by SPSS LOGISTIC REGRESSION.

Interpreting Unstandardized
Logistic Regression Coefficients

Figure 1.4 provided the results of a bivariate logistic regression analysis. From Figure 1.4, we obtained the equation logit(PMRJ5) = .407(EDF5) − 5.487. When EDF5 is at its maximum observed value, 29, this becomes logit(PMRJ5) = .407(29) − 5.487 = 6.316; and if EDF5 has its minimum observed value, 8, it becomes logit(PMRJ5) = .407(8) − 5.487 = −2.231. Translating the logits into probabilities, the probability of marijuana use for individuals whose score on the exposure scale is 29 becomes $e^{6.316}/(1 + e^{6.316})$ = .998, and for individuals whose score on the exposure scale is 8, it becomes $e^{-2.231}/(1 + e^{-2.231})$ = .097. For individuals with the highest levels of exposure to delinquent friends, marijuana use is almost, but not quite, certain. For individuals with the lowest levels of exposure, the relative frequency of marijuana use is less than 10%, low, but far from indicating that marijuana use never occurs among these individuals. At the mean level of exposure, logit(PMRJ5) = .407(12) − 5.487 = −.603, and the probability of using marijuana is $e^{-.603}/(1 + e^{-.603})$ = .354, which is approximately equal to the unconditional probability of marijuana use (P = .357) for this sample of 16-year-olds.

Like the linear regression coefficient, the logistic regression coefficient can be interpreted as the change in the dependent variable, logit(Y), associated with a one-unit change in the independent variable. The change in P(Y = 1), however, is not a linear function of the independent variables. The slope of the curve varies, depending on the value of the independent variables. It is possible to calculate the slope of the curve for different pairs of points by examining the change in P(Y = 1) between those points. For example, going from EDF5 = 8 to EDF5 = 9 results in a change in probability from .097 to .101, indicating a slope of .004. A change from EDF5 = 28 to EDF5 = 29 is associated with a change from P(Y = 1) = .997 to .998, or a slope of about .001. Between EDF5 = 12 and EDF5 = 13, the probability of marijuana use changes from .354 to .451, a slope of .097, many times larger than the changes resulting from one-unit changes at very high or very low values of EDF5.

The interpretation of the logistic regression coefficient is similar in models with several independent variables. The equation for the relationship between prevalence of marijuana use and the predictors in Figure 3.1 is

$$\text{logit(PMRJ5)} = .407(\text{EDF5}) - .118(\text{BELIEF4}) - 1.514(\text{SEX})$$
$$+ .245(\text{BLACK}) + .772(\text{OTHER}) - 1.749,$$

where BLACK and OTHER are the descriptive labels associated with ETHN(1) and ETHN(2) in Figure 3.1. Turning to the individual coefficients, each one-unit increase in EDF5 is associated with an increase of .407 in logit(PMRJ5). Each one-unit increase in BELIEF4 is associated with a decrease of .118 in logit(PMRJ5). Being male reduces the logit of PMRJ5 by 1.514 (remember, in this sample, males have lower marijuana use than females). The effects of ethnicity are not statistically significant.[15]

Predictions for individual cases may be obtained by replacing the variables in the equation with their values for specific cases. For example, for an African American female (BLACK = 1, OTHER = 0) with strong beliefs that it is wrong to violate the law (BELIEF4 = 25) and low levels of exposure to delinquent friends (EDF5 = 10), logit(PMRJ5) = .407(10) − .118(25) − 1.514(0) + .245(1) + .772(0) − 1.749 = −.384. This corresponds to a probability of marijuana use of $e^{-.384}/(1 + e^{-.384}) = .405$. Alternatively, for a non-Hispanic European American male (BLACK = 0, OTHER = 0) with moderate levels of both belief that it is wrong to violate the law (BELIEF4 = 20) and exposure to delinquent friends (EDF5 = 15), the equation becomes logit(PMRJ5) = .407(15) − .118(20) − 1.514(1) + .245(0) + .772(0) − 1.749 = .482. This corresponds to a probability of marijuana use of $e^{.482}/(1 + e^{.482}) = .618$.

Substantive Significance and Standardized Coefficients

What does a one-unit increase in exposure to delinquent friends really mean? Because exposure items are measured on a 5-point scale and belief items are measured on a 4-point scale, and because the number of items is larger for exposure (8) than for belief (7), is a one-unit increase in belief really equivalent to a one-unit increase in exposure? Should we regard a one-unit change in belief (which has, in principle, a range of 7 to 28) as equivalent to a one-unit change in gender (which has a range of only one unit, 0 to 1)? These questions could be asked in the context of either linear regression, with frequency of marijuana use as a dependent variable, or logistic regression, with prevalence of marijuana use as a dependent variable. When independent variables are measured in different units or on different scales and we want to compare the strength of the relationship between the dependent variable and different independent variables, we often use standardized regression coefficients in linear regression analysis. For the same reasons, we may want to consider using standardized coefficients in logistic regression analysis.

A standardized coefficient is a coefficient that has been calculated for variables measured in standard deviation units. A standardized coefficient indicates how many standard deviations of change in a dependent variable are associated with a 1 standard deviation increase in the independent variable. In linear regression, a standardized coefficient between a dependent variable Y and an independent variable X, b_{YX}^*, may be calculated from the unstandardized coefficient between Y and X, b_{YX}, and the standard deviations of the two variables, s_Y and s_X: $b_{YX}^* = (b_{YX})(s_X)/(s_Y)$. Alternatively, standardizing both X and Y prior to regression by subtracting their respective means and dividing by their respective standard deviations, to obtain $Z_Y = (Y - \overline{Y})/s_Y$ and $Z_X = (X - \overline{X})/s_X$, produces a standardized regression coefficient between Y and X.

For a variable that is approximately normally distributed, 99.9865% of all cases will lie in a range of 6 standard deviations (3 standard deviations on either side of the mean), and 99.999999713% will lie within a range of 10 standard deviations. Thus, a 1 standard deviation change in an independent variable typically means a change of about one eighth of the range of its possible values (one sixth in a small sample, one tenth in a very large sample). According to Chebycheff's inequality theorem (Bohrnstedt & Knoke, 1988, pp. 141-144), for any distribution, even for a very nonnormal distribution, at least 93.75% of all cases will lie within 8 standard deviations of the mean, and 96% within 10 standard deviations. Thus a change of 1 standard deviation seems intuitively to be a large enough change that its effect should be felt (if the independent variable has any impact on the dependent variable), but not so large that a trivial relationship should appear to be substantial, even in a distribution that departs considerably from a normal distribution. By measuring the relationship of all of the independent variables to the dependent variables in common units (standard deviations, or about one eighth of their range), the relative impact on the dependent variable of independent variables measured in different units can be directly compared.

In logistic regression analysis, the calculation of standardized coefficients is complicated by the fact that it is not the value of Y, but the probability that Y has one or the other of its possible values, that is predicted by the logistic regression equation. The actual dependent variable in logistic regression is not Y, but logit(Y), whose observed values of logit(0) = $-\infty$ and logit($+\infty$) = $+\infty$ do not permit the calculation of means or standard deviations. Although we cannot directly calculate the standard deviation for the observed values of logit(Y), we can calculate the standard deviation indirectly, using the predicted values of logit(Y) and the explained variance, R^2. Recall from Chapter 2 that R^2 = SSR/SST. Dividing both the numerator and the divisor by N ($N-1$ for a sample), we get R^2 = SSR/SST = (SSR/N)/(SST/N) = $s_{\hat{Y}}^2/s_Y^2$. Rearranging this equation to solve for s_Y^2 produces the equation $s_Y^2 = s_{\hat{Y}}^2/R^2$, and substituting logit(Y) for Y and logit($\hat{Y}$) for $\hat{Y}$, we are able to calculate the variance of logit(Y) based on the standard deviation of the predicted values of logit(Y) and the explained variance. Because the standard deviation is the square root of the variance, we can estimate standardized logistic regression coefficients as

$$b_{YX}^* = (b_{YX})(s_X)/\sqrt{s_{\text{logit}(\hat{Y})}^2/R^2} = (b_{YX})(s_X)(R)/s_{\text{logit}(\hat{Y})} \qquad (3.1)$$

where b_{YX}^* is the standardized logistic regression coefficient, b_{YX} is the unstandardized logistic regression coefficient, s_X is the standard deviation of the independent variable X, $s_{\text{logit}(\hat{Y})}^2$ is the variance of logit($\hat{Y}$) [in other words, the variance of the estimated values of logit(Y)], $s_{\text{logit}(\hat{Y})}$ is the standard deviation of logit($\hat{Y}$), and R^2 is the coefficient of determination.

In order to calculate standardized logistic regression coefficients with existing SAS and SPSS software, the following steps are necessary:

1. b: Calculate the logistic regression model to obtain the unstandardized logistic regression coefficient, b. Save the predicted value of Y from the logistic regression model.

2. R: Use the predicted value of Y to calculate R^2, R, η^2, or η (because these measures convey the same information, it does not matter which one you calculate).

3. Use the predicted value of Y to calculate the predicted value of logit(Y), using the equation logit($\hat{Y}$) = ln[$\hat{Y}/(1 - \hat{Y})$].

4. $s_{\text{logit}(\hat{Y})}$: Calculate descriptive statistics for logit($\hat{Y}$), including the standard deviation.

5. s_X: If you have not already done so, calculate the standard deviations of all of the independent variables in the equation. Be sure that you calculate

TABLE 3.1

Logistic Regression Analysis Results for Prevalence of Marijuana Use

Dependent Variable	Association/ Predictive Efficiency	Independent Variable	Unstandardized Logistic Regression Coefficient (b)	Standard Error of b	Statistical Significance of b	Standardized Logistic Regression Coefficient
PMRJ5	G_M = 108.257	EDF5	.407	.069	.000	.531
	(p = .000)					
	R_L^2 = .367	BELIEF4	−.118	.060	.048	−.143
	R^2 = .435	SEX (male)	−1.514	.405	.000	−.233
		ETHN			.552	
	λ_p = .488	(black)	.245	.508	.630	.028
		(other)	.772	.745	.300	.059
	τ_p = .604	Intercept	−1.749	2.028	.388	_____

them only for the cases actually included in the model. (In other words, use listwise deletion of missing data when you calculate the descriptive statistics.)

6. Enter b, R (or η), s_X, and $s_{\text{logit}(\hat{Y})}$ into equation 3.1 above to calculate b^*.

The interpretation of the standardized logistic regression coefficient, calculated as $b^* = bs_X R/s_Y$, is straightforward and closely parallels the interpretation of standardized coefficients in linear regression: A 1 standard deviation increase in X produces a b^* standard deviation change in logit(Y). For the model in Figure 1.4, the standard deviation of EDF5, the standard deviation of logit($\hat{Y}$), and η were obtained separately. The standard deviation of EDF5 was 4.24, the standard deviation of logit($\hat{Y}$) was $s_{\text{logit}(\hat{Y})}$ = 1.72, $R = \eta = .5871$, and $b = .4068$. From Equation 3.1, $b^* = (.4068)(4.24)(.5871)/$ 1.72 = .591. In other words, a 1 standard deviation increase in EDF5 is associated with an increase of .591 standard deviations in logit(PMRJ5).

Table 3.1 summarizes the output from SAS PROC LOGISTIC and SPSS LOGISTIC REGRESSION and adds measures of goodness of fit and predictive efficiency. The relationship between the dependent variable and the independent variables is statistically significant: $G = 108.257$ with 5 degrees of freedom, $p = .000$. Measures of the strength of association between the dependent variable and the independent variables, $R_L^2 = .367$ and $R^2 = \eta^2 = .435$ (the latter from Figure 3.1), indicate a moderately strong relationship between the dependent variable and its predictors. The indices

48

of predictive efficiency also indicate a model that predicts well: $\lambda_p = .488$ and $\tau_p = .604$, both statistically significant at $p = .000$. Comparison of Table 3.1 with Figure 3.2 reveals that the standardized coefficients in Table 3.1 do not match the "Standardized Estimate" provided by SAS in Figure 3.2. This is because SAS calculates the standardized estimate of the logistic regression coefficient as $b^*_{SAS} = (b)(s_X)/(\pi/\sqrt{3}) = (b)(s_X)/1.8138$. The quantity $\pi/\sqrt{3}$ is the standard deviation of the standard logistic distribution (just as 1 is the standard deviation of the standard normal distribution). The "standardized" coefficients provided by SAS are really partially, not fully, standardized; they do not take the actual distribution of Y or logit(Y) into account, but divide by the same constant regardless of the distribution of Y. Another alternative is to standardize only the independent variables. Both the SAS and independents-only approach to partial standardization produce the same ranking of effects of independent variables on the dependent variable as full standardization, but limited experience suggests that they are more likely than the fully standardized coefficient b^* to be greater than 1 or less than −1 even when there are no problems of collinearity or other problems (see Chapter 4). The principal reasons favoring the use of the fully standardized coefficient are (a) construction and interpretation that directly parallel the standardized coefficients in linear regression and, correspondingly, (b) the ability to apply the same standards used to interpret standardized coefficients in linear regression to standardized coefficients in logistic regression.

If we naively evaluate the strength of the relationships of the independent variables to PMRJ5 based on the unstandardized logistic regression coefficients (or, equivalently, based on odds ratios or probabilities), SEX appears to have the strongest effect, followed by EDF5 and BELIEF4. (ETHN is not statistically significant.) Based on the standardized coefficients, however, EDF5 appears to have the strongest effect (.531 in Table 3.1), followed by SEX (−.233), and then BELIEF4 (−.143). In other words, (a) a 1 standard deviation increase in EDF5 is associated with a .531 standard deviation increase in logit(PMRJ5), (b) a 1 standard deviation increase in BELIEF4 is associated with a .143 standard deviation decrease in logit(PMRJ5), and (c) a 1 standard deviation increase (becoming "more male") in SEX is associated with a .418 standard deviation decrease in logit(PMRJ5). Changes in ETHN, which is not statistically significant as a predictor of PMRJ5, are associated with changes of less than one tenth of a standard deviation in logit(PMRJ5).

For SEX and ETHN, a "1 standard deviation increase" is not as intuitively meaningful as the difference between males and females, or between respondents from different ethnic backgrounds, as reflected in the unstandardized logistic regression coefficient. The real utility of the standardized logistic regression coefficient here is to compare the magnitude of the effects of the predictors by converting them to a common scale of measurement. In presenting substantive results, it may make sense to focus on standardized coefficients for unit-less scales like EDF5 and BELIEF4, but unstandardized coefficients for categorical variables like ETHN and SEX (corresponding to realistic differences in ethnicity and gender), and perhaps for variables with natural units of measurement (inches, kilograms, dollars, number of occasions) as well.

Odds Ratios

In the last column of statistics in Figures 3.1 (under "Variables in the Equation") and 3.2 (under the heading "Analysis of Maximum Likelihood Estimates"), the *odds ratio* associated with each coefficient is presented, as Exp(B) in SPSS and as Odds Ratio in SAS. The odds ratio is the number by which we would multiply the odds of being a marijuana user (the probability, divided by 1 minus the probability) for each one-unit increase in the independent variable. An odds ratio greater than 1 indicates that the odds of being a marijuana user increase when the independent variable increases; and an odds ratio of less than 1 indicates that the odds of being a marijuana user decrease when the independent variable increases. For example, a one-unit increase in EDF5 results in a 50.2% increase in the odds of being a marijuana user (the odds of being a marijuana user are multiplied by 1.502). A one-unit increase in BELIEF4 decreases the odds of being a marijuana user by 11.1% (the odds of being a marijuana user are multiplied by .889, which is .111 less than 1).

It is important to emphasize that the odds ratio is not a separate measure of the relationship between the dependent variables and the independent variables. It contains the same information as the logistic regression coefficient or the probability. All that is different is the way in which the information is presented. In particular, the odds ratio cannot take the place of a standardized logistic regression coefficient for evaluating the strength of the influences of the independent variables on the dependent variable, relative to one another, because the odds ratio will provide exactly the same ordering, from strongest to weakest, as the unstandardized logistic regression coefficient, once all of the odds ratios are transformed to be greater than 1 (or all less than 1). The odds ratio provides no additional information; it just provides the same information as the logistic regression coefficient in a different way.

More on Categorical Predictors:
Contrasts and Interpretation

The use of 0s and 1s to represent the different possible values of the variable ETHN in Figure 3.1 and Table 3.1 is called *indicator* coding because it indicates the presence or absence of a categorical attribute. Indicator coding is only one of several ways of treating design variables in logistic regression analysis. One alternative is called *simple* coding in SPSS LOGISTIC REGRESSION. With simple contrasts, logistic regression coefficients for the design variables are identical to the coefficients produced with indicator coding, only the intercept changes.

Another alternative is *deviation* coding, the default option in SPSS LOGISTIC REGRESSION. With deviation coding, the effect of each design variable is compared with the overall effect of the independent variable. This is analogous to comparing the means (not weighted by number of cases) for the three categories in regression or analysis of variance. In logistic regression, the deviation coding measures the deviation of the logit for each group from the average logit for the entire sample. With deviation coding, the reference category no longer has an arbitrary coefficient of zero. Instead, its coefficient is equal to the negative of the sum of the coefficients for the other categories. If computer time is more expensive than human labor, calculation of the omitted coefficient by hand may be reasonable. In a personal computer or free computing environment, however, it makes more sense to calculate two models, with different reference categories, to obtain not only the estimates for the coefficients of all three categories but also the standard error and statistical significance of the otherwise omitted category.[16] The model with deviation coding for ETHN is summarized in Table 3.2.

Other than the changes in the individual coefficients for ETHN, the use of a different coding scheme for the indicator variables has not changed the results of the analysis. The order of the ethnic groups is the same as in Figure 3.1 (non-Hispanic European Americans have the lowest probability and Others the highest probability of marijuana use) but, unlike Table 3.1, Table 3.2 adds the information that being African American would be, on the whole, *negatively* associated with the probability of marijuana use if the effect of ethnicity were statistically significant. The coefficients for indicator contrasts in Table 3.1 can be reconstructed from Table 3.2 by

TABLE 3.2

Logistic Regression for Prevalence of Marijuana Use:
Deviation Coding of Ethnicity

Dependent Variable	Association/ Predictive Efficiency	Independent Variable	Unstandardized Logistic Regression Coefficient (b)	Standard Error of b	Statistical Significance of b	Standardized Logistic Regression Coefficient
PMRJ5	$G_M = 108.257$	EDF5	.407	.069	.000	.531
	($p = .000$)					
	$R_L^2 = .367$	BELIEF4	−.118	.060	.048	−.143
	$R^2 = .435$	SEX (male)	−1.514	.405	.000	−.233
		ETHN			.552	
		(white)	−.339	.391	.289	−.044
	$\lambda_p = .488$	(black)	−.094	.319	.810	−.011
		(other)	.433	.388	.388	.033
	$\tau_p = .604$	Intercept	−1.410	2.042	.490	———

subtracting −.339, the coefficient for non-Hispanic Europeans, from each of the other coefficients in Table 3.2, but the coefficients in Table 3.2 cannot similarly be reconstructed from the data in Table 3.1. Deviation coding thus gives us a little more information than indicator coding.

Other contrasts available for logistic regression analysis in SPSS include Helmert, reverse Helmert, polynomial, repeated, and special contrasts. Helmert, reverse Helmert, orthogonal, and repeated contrasts are appropriate for testing whether the effects of different categories of an ordinal predictor are consistent with the ordering of the categories. Polynomial contrasts test for linear and nonlinear effects. The use of different contrasts for ordinal variables has no effect on the model fit or on the statistical significance of the categorical ordinal variable. The results may, however, suggest an appropriate recoding of the variable to take advantage of any apparent linearity, monotonicity, or any natural breaks between categories. The simplest ordinal contrast is the repeated (SPSS) or profile (SAS) contrast, in which each category of the independent variable except the first (the reference category) is compared with the previous category. By examining the coefficients for the categories, it is possible to see whether a monotonic or linear relationship exists between the independent variable and the dependent variable. If there is a nonsystematic pattern of positive and negative coefficients, a nonlinear, nonmonotonic relationship is indicated, and the independent variable is best treated as though it were nominal rather than ordinal.

Whenever design variables are used to represent the effect of a single *nominal* variable, it is important that the design variables be treated *as a group*, rather than as individual variables. The statistical significance of the individual design variables should be considered only if the design variables as a group have a statistically significant effect on the dependent variable. The statistical significance of the individual design variables should be interpreted as whether the effects of being in a certain category is statistically significantly different from being in the reference category (for indicator coding) or from the average effect of the categorical variable (in deviation coding), *given that the categorical variable has a statistically significant effect to begin with.* In SPSS, a test of the effect of the statistical significance of the nominal variable (all the design variables taken together) is provided. In SAS, a similar test can be obtained by comparing the model with and without the nominal variable, that is, with and without all of the indicator variables used to represent ethnicity, using stepwise procedures. For ordinal contrasts, the overall statistical significance of the design variable indicates only whether the categorical variable, *treated as a nominal variable*, has a statistically significant effect on the dependent variable. For ordinal contrasts, the statistical significance of the individual coefficients may provide important information about the form of the relationship between the categorical predictor and the dependent variable, even when the categorical variable does not appear to have a statistically significant effect on the dependent variable.

Interaction Effects

In some statistical software, one need only specify the interaction term to be included and the software calculates the interaction term, includes it in the equation, and provides information about its statistical significance and the strength of its relationship to the dependent variable. In other software packages, it is necessary to calculate the interaction term (or terms, if the interaction involves a nominal variable with more than two categories) separately and add it (or them) to the model. The only complication here is when the interaction involves a nominal variable with more than two categories, in which case it may be necessary to compare the model with and without all of the interaction terms to determine whether the interaction is statistically and substantively significant. In linear regres-

of the use of stepwise techniques in exploratory research is provided by Wofford, Elliott, and Menard (1994). Wofford et al. studied the continuity of domestic violence in a national probability sample of young men and women, 18-27 years old. Twenty-six predictors, based on the domestic violence literature, were included in their analysis. As part of the study, respondents who had reported being victims or perpetrators of domestic violence in 1984 were reinterviewed in 1987 to see whether the domestic violence had continued or been suspended since the 1984 interview. A total of 108 women (out of 807 in the original sample) reported being victims of domestic violence in 1983 and were reinterviewed in 1986. Wofford et al. constructed a logistic regression model that included all 26 predictors. Because theory in this area was not well developed, and because the number of cases was small relative to the number of explanatory variables suggested in the literature, stepwise logistic regression was used.

Backward elimination rather than forward inclusion was selected as the method of stepwise regression. In some cases, a variable may appear to have a statistically significant effect only when another variable is controlled or held constant. This is called a *suppressor* effect (Agresti & Finlay, 1986, pp. 304-305). One disadvantage to forward inclusion as a method for stepwise regression is the possible exclusion of variables involved in suppressor effects. With backward elimination, because both variables will already be in the model, there is less risk of failing to find a relationship when one exists. Usually, the results of backward elimination and forward inclusion methods of stepwise linear regression will be the same, but when they differ, backward elimination may uncover relationships missed by forward inclusion.

To further prevent the failure to find a relationship when one exists, the usual .05 criterion for statistical significance should probably be relaxed. Bendel and Afifi (1977), based on their studies of forward stepwise regression, suggested that .05 is too low and often excludes important variables from the model. Instead, they recommended that the statistical significance criterion for inclusion be set in a range from .15 to .20. This results in an increased risk of rejecting the null hypothesis when it is true (finding a relationship that is not really there) but a lower risk of failing to reject the null hypothesis when it is false (not finding a relationship that really is there). In exploratory research, as opposed to theory testing, there tends to be a greater emphasis on finding good predictors than on eliminating bad ones. Wofford et al. examined three models: a full model with all of the variables in the logistic regression equation, a reduced model with

TABLE 3.4
Continuity of Marital Violence Victimization (Women)

$N = 108$	Model 1: All Variables Included	Model 2: Maximum $p = .100$	Model 3: Maximum $p = .050$
Model chi-square G_M (degrees of freedom)	30.254 (28 df)	21.284 (7 df)	4.472 (1 df)
Statistical significance of G_M	.351	.003	.034
Deviation chi-square D_M (degrees of freedom)	119.429 (79 df)	128.298 (100 df)	145.211 (106 df)
Statistical significance of D_M	.002	.029	.007
Change in G_M from previous model	———	8.870 (21 df)	16.812 (6 df)
Statistical significance of change in G_M from previous model	———	.99	.010
R_L^2	.202	.150	.030
τ_p	.481	.509	.145

Individual Predictor Results for Model 2

Independent Variables	b	Standard Error	p (Based on Likelihood Ratio Statistic)
Welfare recipient	1.88	.95	.03
Social class background	−0.03	.02	.05
Prior minor assault	1.24	.53	.02
Prior felony assault	−1.07	.62	.08
Witnessed parental violence	−1.70	.64	.00
Frequency of serious violence in relationship	0.12	.06	.05
Sought professional assistance	0.88	.53	.09

all variables for which $p > .10$ eliminated (in practice, this was the same as using a .15 or .20 cutoff), and a further reduced model with all variables for which $p > .05$ eliminated. Results are presented in Table 3.4.

The first part of Table 3.4 compares the three models. For the full model, G_M is not statistically significant, indicating that the predictor variables contribute no more than chance to the explanation of the dependent variable. One reason for the failure of the model chi-square to attain statistical significance is the small sample size; another may be the large number of variables included in the model. Model 2 ($p < .10$) has a smaller but statistically significant G_M, as does Model 3 ($p < .05$), with only one predictor in the model. The change in the model chi-square (or equivalently the change in the deviation chi-square) from Model 1 to Model 2 and from Model 1 to Model 3 is not statistically significant; however, the change in G_M from Model 2 to Model 3 is statistically significant at the .01 level. For the full model, R_L^2 is .20; for Model 2, it decreases to .15; and for Model 3, it is only .03. Model 1 has a τ_p of .48; τ_p actually increases to .51 for Model 2, but for Model 3 it is only .14.

Model 2 was selected for further analysis because (a) G_M was statistically significant for the reduced models but not the full model, (b) Model 2 provided a statistically significantly better fit than Model 3 and did not fit statistically significantly worse than the full model, and (c) the changes in R_L^2 and τ_p were relatively small (and in opposite directions) for Model 1 compared with Model 2, but R_L^2 and τ_p were much lower for Model 3 than for Model 2. The results of Model 2 are presented in the second half of Table 3.4. Substantively, they indicate that women who are welfare recipients, from a higher social class background, who have committed minor assaults but who have not committed felony assaults, who have not witnessed parental violence, who have experienced higher frequencies of serious violence in the relationship, and who have sought professional assistance are more likely to experience continuity rather than suspension of domestic violence. Full discussion of the substantive results is left to Wofford et al. (1994).

Several methodological points regarding the use of stepwise logistic regression are illustrated in Figure 2.1. Probably the most important methodological point is that these results must be regarded as very tentative and inconclusive. This is a search for plausible predictors, not a convincing test of any theory. Second, an important element of the stepwise procedure is the comparison of the full and reduced models. As suggested by Bendel and Afifi (1977), the .05 criterion for inclusion appears to be too severe; based on the comparisons of goodness of fit and predictive efficiency statistics, more reasonable results are obtained with a more liberal cutoff point for statistical significance. Third and finally, the variables identified in Model 2 are good candidates for use in the prediction of domestic violence, but some may as easily be effects (e.g., seeking professional assistance) as causes. Further development and testing of theory may be based on these results but would require replication with other data and explanation (preferably in the form of a clear theoretical justification) of why these variables appear as predictors of continuity of domestic violence.

58

4. AN INTRODUCTION TO
LOGISTIC REGRESSION DIAGNOSTICS

When the assumptions of logistic regression analysis are violated, calculation of a logistic regression model may result in one of three problematic effects: biased coefficients, inefficient estimates, or invalid statistical inferences. *Bias* refers to the existence of a systematic tendency for the estimated logistic regression coefficients to be too high or too low, too far from zero or too close to zero, compared to the true values of the coefficients. *Inefficiency* refers to the tendency of the coefficients to have large standard errors relative to the size of the coefficient. This makes it more difficult to reject the null hypothesis (the hypothesis that there is no relationship between the dependent variable and the independent variable) even when the null hypothesis is false. *Invalid statistical inference* refers to the situation in which the calculated statistical significance of the logistic regression coefficients is inaccurate.

In addition, the presence of cases with unusually high or low values on the independent variables (not on the dependent variable, which has only two values) or with unusual combinations of values on the dependent and independent variables may exert a disproportionate influence on the estimated parameters. Such cases are often called *outliers* if they have unusually high or low values on one variable or an unusual combination of values on two or more variables and *influential* cases if they exert a disproportionate influence on the estimates of the logistic regression coefficients. This chapter focuses on the consequences of violations of logistic regression assumptions and on methods for detecting and correcting violations of logistic regression assumptions. Also considered here are methods for detecting outliers and influential cases in logistic regression analysis, as well as alternative approaches for dealing with those cases.

Specification Error

The first and most important assumption in both linear and logistic regression analysis is that the model is correctly specified. Correct specification has two components: (a) the functional form of the model is correct and (b) the model includes all relevant independent variables and no irrelevant independent variables. Misspecification may result in biased logistic regression coefficients, coefficients that are systematically overestimated or underestimated. In Chapter 1, we saw that the linear regression model, when applied to a dichotomous dependent variable, appeared to be misspecified. This led us to examine the logistic regression model. It may nonetheless be the case that the logistic regression model, with logit(Y) as the dependent variable and with a *linear* combination of independent variables, is incorrect in its functional form. First, logit(Y) may be equal to a *nonlinear* combination of the independent variables. Second, the relationship among some or all of the independent variables may be *multiplicative* or *interactive*, rather than additive.

Misspecification as a result of using the logistic function, as opposed to a different S-shaped function, is less likely to be a problem. Aldrich and Nelson (1984) demonstrated that logit models (based on the logistic distribution) and probit models (based on the normal distribution) produce highly similar results. Hosmer and Lemeshow (1989, p. 168) note that logistic regression models are highly flexible and produce very similar results to other models in the range of probabilities between .2 and .8. There is usually little theoretical basis for preferring an alternative model.

Omitting Relevant Variables
and Including Irrelevant Variables

Including one or more irrelevant variables has the effect of increasing the standard error of the parameter estimates, that is, of reducing the efficiency of the estimates, without biasing the coefficients. The degree to which the standard errors are inflated depends on the magnitude of the correlation between the irrelevant included variable and the other variables in the model. If the irrelevant included variable is completely uncorrelated with the other variables in the equation, the standard errors may not be inflated at all, but this condition is extremely unlikely in practice.

Omitting relevant variables from the equation in logistic regression results in biased coefficients for the independent variables, to the extent that the omitted variable is correlated with the independent variables in the logistic regression equation. As in linear regression (Berry & Feldman, 1985), the direction of the bias depends on the parameter for the excluded variable, the direction of the effect of the excluded variable on the dependent variable, and the direction of the relationship between excluded and included variables. The magnitude of the bias depends on the strength of the relationship between the included and excluded variables. If the excluded variable is completely uncorrelated with the included variables, the coefficients may be unbiased, but in practice this is unlikely to occur. Bias is generally regarded as a more serious problem than inefficiency, but a small amount of bias may be preferable to massive inefficiency.

Omitted variable bias may occur because available theories have failed to identify all of the relevant predictors or causes of a dependent variable, or because theoretically relevant variables have been omitted. The pattern characteristic of omitted variable bias may also occur if the functional form of the model is misspecified. A linear specification of a nonlinear model may be computationally equivalent to the omission of a variable representing a nonlinear component of the relationship between the dependent variable and an independent variable. An additive specification of a nonadditive model may be equivalent to the omission of a variable, specifically a variable constructed as the interaction of two other variables, from the model. When the omitted variable is neither a nonlinear term nor an interaction term, only theory (or perhaps a disappointingly low R_L^2) offers much hope of identifying and remedying the problem. When the excluded variable is really a nonlinear term or an interaction term, a function of variables already in the equation, then the detection and correction of the problem can be considerably easier.

Nonlinearity in the Logit

In a linear regression model, the change in the dependent variable associated with a one-unit change in the independent variable is constant, equal to the regression coefficient for the independent variable. If the change in Y for a one-unit change in X depends on the value of X (as it does when Y is a dichotomous variable), the relationship is nonlinear. Correspondingly, when logit(Y) is the dependent variable, if the change in logit(Y) for a one-unit change in X is constant and does not depend on the value of X, we say that the logistic regression model has a linear form, or that the relationship is *linear in the logit*, and the change in logit(Y) for a one-unit change in X is equal to the logistic regression coefficient. If the relationship is not linear in the logit, the change in logit(Y) for a one-unit change in X is not constant, but depends on the value of X.

There are several possible techniques for detecting nonlinearity in the relationship between the dependent variable, logit(Y), and each of the independent variables (Hosmer & Lemeshow, 1989, pp. 88-91). One is to treat each of the independent variables as a categorical variable and use an orthogonal polynomial contrast to test for linear, quadratic, cubic, and higher order effects either in bivariate logistic regression or in a multiple logistic regression model. If the independent variable has a large number of categories (e.g., 20), the standard errors tend to be large, and neither the linear nor any of the nonlinear effects may appear to be statistically significant, even when a statistically significant linear effect exists. A second possibility is to use the Box-Tidwell transformation described by Hosmer and Lemeshow (1989, p. 90). This involves adding a term of the form $(X)\ln(X)$ to the equation. If the coefficient for this variable is statistically significant, there is evidence of nonlinearity in the relationship between logit(Y) and X. Hosmer and Lemeshow note that this procedure is not sensitive to small departures from linearity. Additionally, this procedure does not specify the precise form of the nonlinearity. If the relationship is nonlinear, further investigation is necessary to determine the pattern of the nonlinearity.

A third procedure suggested by Hosmer and Lemeshow is to aggregate cases into groups defined by the values of the independent variable X, calculate the mean of the dependent variable Y for each group, then take the logit of the mean of Y for each group and plot it against the value of the independent variable. For each value i of the independent variable X, the mean of Y is the probability $P(Y = 1 \mid X = i)$. One problem with this procedure arises if, for any value of X, Y is always either 1 or 0. If it is, then we cannot calculate logit($\overline{Y}$), which would be equal to $\pm\infty$, either infinitely large or infinitely small. It may be possible to overcome this problem by grouping adjacent categories with similar but unequal probabilities. This could conceal some of the nonlinearity in the relationship, however. Another possible option would be to assign an arbitrarily large mean (e.g., .99) to groups with a mean of 1, and an arbitrarily small mean (.01) to groups with a mean of 0, in order to implement this method. An important advantage to this method is that, like graphical techniques generally, it helps identify the pattern of the nonlinearity. In addition, examination of the plot may help identify cases with unusual values on the independent variable or combinations of values on the dependent and independent variables.

Table 4.1 presents the results of a Box-Tidwell test for nonlinearity. In the first part of Table 4.1, the two nonlinear terms BTEDF = (EDF5)ln(EDF5) and BTBEL = (BELIEF4)ln(BELIEF4) are added to the model. Taken together, the effects of the two nonlinear interaction terms are statistically significant (change in G_M = 15.066 with 2 degrees of freedom; p = .005), but based on the G_X, the likelihood ratio statistic for each of the nonlinear terms, only BTBEL is statistically significant (G_X = 9.932 with 1 degree of freedom, p = .002 for BTBEL; G_X = 2.226 with 1 degree of freedom, p = .136 for BTEDF). When BTEDF is removed from the model, the coefficient for BTBEL is still statistically significant (G_X = 12.839, 1 degree of freedom, p = .000), and inclusion of BTBEL in the equation increases R_L^2 by .034, or 3.4%.

Figure 4.1 shows why the relationship between PMRJ5 and BELIEF4 appears to be nonlinear. The mean of PMRJ5 was calculated for each value

TABLE 4.1

Box-Tidwell Tests for Nonlinearity

Dependent Variable	Association/ Predictive Efficiency	Independent Variable	Unstandardized Logistic Regression Coefficient (b)	Standard Error of b	Statistical Significance of b (Based on likelihood Ratio G_X)	Standardized Logistic Regression Coefficient
PMRJ5	G_M=123.322	EDF5	2.415	1.191	.076	3.073
n = 227	(p = .000)					
		BELIEF4	3.620	1.421	.002	−4.272
	$R_L^2 = .418$					
		SEX (male)	−1.660	.428	.000	−.249
	Change in					
	G_M from	ETHN			.516	
	base model	(black)	.333	.536	.535*	.036
	(Table 3.1) =	(other)	.859	.817	.293*	.063
	15.065					
	(p = .001)	BTEDF	−.551	.325	.136	−1.620
	Change in	BTBEL	−.891	.340	.002	−4.172
	R_L^2 from					
	base					
	model =	Intercept	−29.210	8.733	.001*	———
	.051					

Continued

of BELIEF4 and, because the mean of PMRJ5 was either 0 or 1 for several values of BELIEF4, values of 1 were recoded as .99 and values of 0 were recoded as .01. Next, the logit of each mean was taken and plotted in Figure 4.1 against the values of BELIEF4. In the lower left quadrant of the plot there are two outliers, respondents with very weak mean beliefs that it is wrong to violate the law, but who report no marijuana use. Each mean, as it turns out, is based on a single case. Except for these two cases, the plot does not appear to depart substantially from linearity. The second half of Table 4.1 confirms this assessment. With the two cases deleted from the analysis, BTEDF and BTBEL, separately and in combination, have no statistically significant effect on the logit of PMRJ5. Whether these cases should be deleted or retained will receive further consideration below.

TABLE 4.1

Continued

Dependent Variable	Association/ Predictive Efficiency	Independent Variable	Unstandardized Logistic Regression Coefficient (b)	Standard Error of b	Statistical Significance of b (Based on likelihood Ratio G_X)	Standardized Logistic Regression Coefficient
PMRJ5	$G_M =$	EDF5	2.411	1.184	.076	3.012
$n = 225$	125.195					
	$(p = .000)$	BELIEF4	1.380	2.812	.644	−1.503
	$R_L^2 = .427$					
		SEX (male)	−1.676	.436	.000	−.249
	Change in					
	G_M from	ETHN	.530			
	base model	(black)	.342	.534	.522*	.036
	$(n = 225) =$	(other)	.823	.814	.312*	.060
	2.561					
	$(p = .278)$	BTEDF	−.555	.323	.137	−2.551
	Change in	BTBEL	−.371	.660	.599	−1.716
	R_L^2 from					
	base	Intercept	−15.541	17.134	.364*	———
	model					
	$(n = 225) =$					
	.009					

*Statistical significance based on the likelihood ratio statistic is not available for individual categories of categorical independent variables or for the intercept; for these, the Wald statistic is used to determine statistical significance.

Nonadditivity

Nonlinearity occurs when the change in the dependent variable associated with a one-unit change in an independent variable depends on the value of the independent variable. Nonadditivity occurs when the change in the dependent variable associated with a one-unit change in an independent variable depends on the value of one of the other independent variables. For example, a one-unit change in exposure to delinquent friends may produce a larger change in the frequency or prevalence of marijuana

64

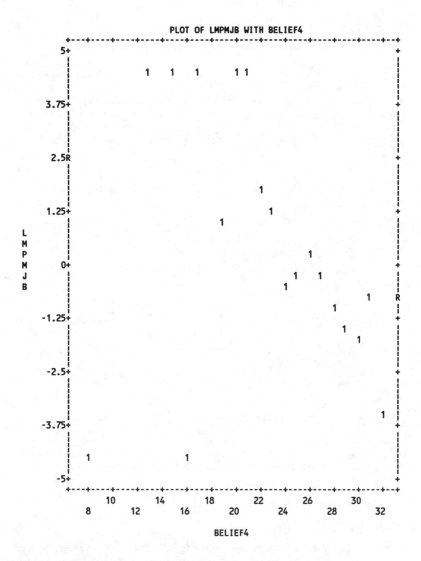

Figure 4.1. Logistic Regression Diagnostics: Test for Nonlinearity

LMPMJB = logit of mean of prevalence of marijuana use for each value of BELIEF4
BELIEF4 = belief that it is wrong to violate the law

use for individuals with weak to moderate beliefs that it is wrong to violate
the law (who may be more susceptible to peer influence) than in individuals

who strongly believe that it is wrong to violate the law (who may be less susceptible to peer influence). Detection of nonadditivity is not as straightforward as detection of nonlinearity in either linear or logistic regression. Unless theory provides some guidance, one is commonly left with the choice among assuming an additive model, testing for interaction effects that seem intuitively plausible, and testing for all possible interaction effects. This last option is feasible for relatively simple models but becomes progressively more tedious and carries increasingly more risk of capitalizing on random sampling variation as the number of variables in the model increases.

One example of an interaction effect was examined in Table 3.3. The results in Table 3.3 indicated that the effects of exposure and belief on the prevalence of marijuana use were not statistically significantly different for males and females. This is an example of an interaction between a continuous predictor (actually two: EDF5 and BELIEF4) and a dichotomous variable. Other possible patterns would include interactions between two categorical predictors (SEX and ETHN) or between two continuous variables (an interaction between EDF5 and BELIEF4).

Collinearity

Collinearity (or colinearity, or multicollinearity) is a problem that arises when independent variables are correlated with one another. *Perfect collinearity* means that at least one independent variable is a perfect linear combination of the others. If we treated each independent variable in turn as the dependent variable in a model with all of the other independent variables as predictors, perfect collinearity would result in an R^2 of 1 for at least one of the variables. When perfect collinearity exists, it is impossible to obtain a unique estimate of the regression coefficients; any of an infinite number of possible combinations of linear or logistic regression coefficients will work equally well. Perfect collinearity is rare, except as an oversight: The inclusion of three variables, one of which is the sum of the other two, would be one example.

Less than perfect collinearity is fairly common. Any correlation among the independent variables is indicative of collinearity. As collinearity increases among the independent variables, linear and logistic regression coefficients will be unbiased, and as efficient as they can be (given the relationships among the independent variables), but the standard errors for linear and logistic regression coefficients will tend to be large. More efficient unbiased estimates may not be possible, but the level of efficiency of the estimates may be poor. Low levels of collinearity are not generally problematic, but high levels of collinearity (perhaps corresponding to an R^2 of .80 or more for at least one of the independent variables) may pose problems, and very high levels of collinearity (perhaps corresponding to an R^2 of .90 or more for at least one of the independent variables) almost certainly result in coefficients that are not statistically significant, even though they may be quite large. Collinearity also tends to produce linear and logistic regression coefficients that appear to be unreasonably high: As a rough guideline, standardized logistic or linear regression coefficients greater than 1, or unstandardized logistic regression coefficients greater than 2, should be examined to determine whether collinearity is present.

For linear regression, detection of collinearity is straightforward. Most standard regression routines in widely used software packages provide optional information on the R^2, or some function of the R^2, for each of the independent variables, when it is treated as the dependent variable with all of the other independent variables as predictors. For example, the *Tolerance* statistic, available in SAS PROC REG and in SPSS REGRESSION, is simply $1 - R_X^2$, where R_X^2 is the variance in each independent variable, X, explained by all of the other independent variables. Corresponding to the rough guidelines outlined above, a tolerance of less than .20 is cause for concern; a tolerance of less than .10 almost certainly indicates a serious collinearity problem. Although tolerance is not available in SAS PROC LOGISTIC or SPSS LOGISTIC REGRESSION, it can easily be obtained by calculating a linear regression model using the same dependent and independent variables as you are using in the logistic regression model. Because the concern is with the relationship among the independent variables, the functional form of the model for the dependent variable is irrelevant to the estimation of collinearity.

In Table 4.1, for both of the models with the nonlinear terms BTEDF and BTBEL, the logistic regression coefficients were somewhat high, and the standardized logistic regression coefficients for EDF5, BELIEF4, BTEDF, and BTBEL were all larger than 1. This suggests that there may be a problem of collinearity in the nonlinear model. Table 4.2 presents collinearity statistics, produced by an OLS regression routine, for two models. The first, labeled "Basic Model," is the logistic regression model from Table 3.1. The second, labeled "Nonlinear Model," is the model from the first half of Table 4.1 with the nonlinear terms BTEDF and BTBEL included. In both models, "BLACK" and "OTHER" are design variables for ETHN. For the basic model, all of the tolerances exceed .70, indicating no serious problem of collinearity. For the nonlinear model, Table 4.2 confirms what the standardized coefficients in Table 4.1 suggested: The

TABLE 4.2
Testing for Collinearity

Dependent Variable	Independent Variable	Tolerance: Basic Model	Tolerance: Nonlinear Model
PMRJ5	EDF5	.717	.00249
	BELIEF4	.707	.00148
	SEX (male)	.994	.994
	ETHN		
	(black)	.959	.958
	(other)	.983	.974
	BTEDF	———	.00253
	BTBEL	———	.00147

tolerances for SEX and the two design variables for ETHN remain high, but BTEDF and BTBEL are severely collinear with EDF5 and BELIEF4, as indicated by tolerances less than .01.

The good news about collinearity is that it is easy to detect. The bad news is that there are few acceptable remedies for it. Deleting variables involved in collinearity runs the risk of omitted variable bias. Combining collinear variables into a single scale, for example by factor analysis, suggests that the theory (if any) used in constructing your model or the measurement process used in collecting your data was faulty, casting doubt on any further inferences you may draw from your analysis. *Ridge regression* (Schaefer, 1986) allows the user to produce somewhat more biased but substantially more efficient estimates by increasing the estimated variance of the variables (thereby decreasing the proportion of the variance that is explained). Perhaps the safest strategy is to focus on the combined effects of all of the variables in the model and to recognize the precariousness of any conclusions about individual predictors in the presence of high collinearity. For a more detailed discussion of remedies to collinearity, see Berry and Feldman (1985, pp. 46-50). Briefly, though, there is no really satisfactory solution to high collinearity.

Numerical Problems:
Zero Cells and Complete Separation

When collinearity exists, it does not necessarily indicate that there is anything wrong with the model or the theory underlying the model. Instead, problems arise because of empirical patterns in the data (the high correlation among independent variables). Two related problems, with similar symptoms, are zero cell count and complete separation. Zero cell count occurs when the dependent variable is invariant for one or more values of a categorical independent variable. If, for example, all of the respondents in the "other" category for ethnicity reported using marijuana (or if they all reported not using marijuana), we would have a problem with a zero cell in the contingency table for the relationship between prevalence of marijuana use and ethnicity. The odds of marijuana use for respondents "other" than white and black would be $1/(1-1) = 1/0 = +\infty$, and the logit = ln(odds) would also be $+\infty$, infinitely large. [If the prevalence of marijuana use were zero for this group, the odds would be $0/(1-0) = 0$, and the logit would be $\ln 0 = -\infty$, infinitely small.] When the odds are 0 or 1 for a single individual or case, this is not a problem; when they are 0 or 1 for an entire group of cases, as defined by the value of a categorical independent variable, the result will be a very high estimated standard error for the coefficient associated with that category (including coefficients for which that category serves as a reference category).

The problem of zero cell count applies specifically to categorical variables, and particularly nominal variables. For continuous variables, and for ordinal categorical variables, it is common to have conditional means of 0 or 1 for some values of the independent variables. The reason that this is not a problem is that we assume a certain pattern to the relationship between the dependent variable and the continuous predictor (linear in linear regression, logistic in logistic regression) and use that pattern to "fill in the blanks" in the distribution of the dependent variable over the values of the independent variable. For categorical variables, we are unable to assume such a pattern. Instead, when we find problems of zero cell count for categorical predictors, we must choose among (a) accepting the high standard errors and the uncertainty about the values of the logistic regression coefficients, (b) recoding the categorical independent variable in a meaningful way (either by collapsing categories or by eliminating the problem category) to eliminate the problem of zero cell count, and (c) adding a constant to each cell of the contingency table to eliminate zero cells.

The first option may be acceptable if we are concerned more with the overall relationship between a set of predictors and a dependent variable than with the effects of the individual predictors. The overall fit of the model should be unaffected by the zero cell count. The third option has no serious drawbacks, but Hosmer and Lemeshow (1989, p. 127) suggest that it may not be adequate for complex analyses. The second option results in cruder measurement of the independent variable and may bias the strength of the relationship between the predictor and the dependent variable toward zero. However, if there is a conceptual link between some categories of the independent variable, and if the distribution of the dependent variable across those categories appears similar, this may be a reasonable option. Usually, this will be done during univariate and bivariate screening of the data. A hidden example of this has been followed throughout this monograph to this point: the coding of ethnicity. In the original survey, ethnicity was divided into six categories: non-Hispanic European American, African American, Hispanic American, Native American, Asian American, and other. The last four categories were collapsed into a single category, "other," because of the small number of cases. Had they been retained in their original form, problems of zero cell count would have plagued the analyses.

If you are too successful in predicting the dependent variable with a set of predictors, you have the problem of complete separation. Both the logistic regression coefficients and their standard errors will tend to be extremely large. The dependent variable will be perfectly predicted: $G_M = D_0$, $D_M = 0$, $R_L^2 = 1$. If separation is less than complete (sometimes called *quasicomplete* separation), logistic regression coefficients and their standard errors will still be extremely large. An example of quasicomplete separation is given in Figure 4.2, based on artificially constructed data. If complete separation occurs in a bivariate logistic relationship, the logistic regression model cannot be calculated. Although there is nothing intrinsically wrong with complete separation (after all, perfect prediction is what we are trying to achieve), as a practical matter it should arouse our suspicions, as it almost never occurs in real-world research. Complete or quasicomplete separation may instead indicate problems in the data or the analysis; for example, having almost as many variables as there are cases to be analyzed.

Collinearity, zero cells, and complete separation have the common symptom of very large standard errors and often, but not always, large coefficients as well (Hosmer & Lemeshow, 1989). All, therefore, result in inefficient estimation of the parameters in the model. None, however, is known to result in biased parameters or in inaccurate (as opposed to inefficient) inferences. Problems with zero cell counts can be averted by careful univariate and bivariate analysis before logistic regression is used. Complete separation may either indicate an error that needs to be corrected or a brilliant breakthrough in theory and analysis. (Congratulations!) Most likely, it indicates a problem. Collinearity is the most bothersome of the

```
logistic regression true with p2/method=enter/classplot.

   Total number of cases:      40 (Unweighted)
   Number rejected because of missing data:  0
   Number of cases included in the analysis: 40
Dependent Variable Encoding:  Original     Internal
                              Value        Value
                                0            0
                                1            1

Dependent Variable..   TRUE
Beginning Block Number  0.  Initial Log Likelihood Function -2 Log Likelihood   55.451774
* Constant is included in the model.
Beginning Block Number  1.  Method: Enter
Variable(s) Entered on Step Number  1..        P2
Estimation terminated at iteration number 9 because Log Likelihood decreased by less than .01 percent.

                      Chi-Square    df Significance
-2 Log Likelihood        13.003     38    .9999
Model Chi-Square         42.448      1    .0000
Improvement              42.448      1    .0000
Goodness of Fit          20.000     38    .9928

Classification Table for TRUE
                    Predicted
                  0        1     Percent Correct
                  0 |  1
Observed      +-------+-------+
   0    0     |  19   |   1   |    95.00%
              +-------+-------+
   1    1     |   1   |  19   |    95.00%
              +-------+-------+
                  Overall        95.00%

--------------------- Variables in the Equation ----------------------

Variable        B        S.E.     Wald      df     Sig      R     Exp(B)

P2         219.7199    74.5348   8.6900      1    .0032   .3473  2.65E+95
Constant  -109.860     37.2749   8.6865      1    .0032

              Observed Groups and Predicted Probabilities

       16 +                                                         +
          |                                                         |
          |                                                         |
    F     |                                                         |
    R  12 +                                                         +
    E     |                                                         |
    Q     |0    1                                        1      1|
    U     |0    0                                        1      1|
    E   8 +0    0                                        1      1+
    N     |0    0                                        1      1|
    C     |0    0                                        1      1|
    Y     |0    0                                        1      1|
        4 +0    0                                        1      1+
          |0    0                                        1      1|
          |0    0                                        1      1|
          |0    0                                        0      1|
Predicted -------------+-------------+-------------+-------------+
  Prob:   0           .25           .5           .75            1
  Group:  000000000000000000000000000000001111111111111111111111111111111111

          Predicted Probability is of Membership for 1
          Symbols: 0 - 0
                   1 - 1
          Each Symbol Represents 1 Case.
```

Figure 4.2. Quasicomplete Separation

three problems, because it indicates a flaw in the theory, a flaw in the operationalization of the theory, or a problem in the empirical data that confounds the testing of the theory, insofar as the theory is concerned with the effects of individual predictors rather than with the combined effect of a set of predictors. Like zero cell counts, collinearity can be detected (with the help of a good multiple regression package) before logistic regression analysis begins. What to do about it if it is detected is problematic, more art than science.

Analysis of Residuals[17]

In linear regression, the residual is commonly denoted e, and $e_j = Y_j - \hat{Y}_j$ is the difference between the observed and predicted values of Y for a given case, j. This should be distinguished from the error of prediction, denoted ϵ_j, which represents the difference between the true value of Y_j in the population (a value that may be different from the observed value of Y in the sample, for example, as a result of measurement error) and the estimated value of Y_j, $\hat{Y}_j$ (Berry, 1993). In linear regression, certain assumptions about errors (zero mean, constant variance or homoscedasticity, normal distribution, no correlation of error terms with one another, no correlation of error terms with independent variables) are necessary if we are to draw statistical inferences from a sample to a larger population. These assumptions may sometimes be tested by using the residuals, e_j, as estimates of the errors, ϵ_j. Violations of some assumptions (zero mean, normal distribution) may have relatively minor consequences. Violation of others is more problematic. Heteroscedasticity inflates standard errors and renders tests of statistical significance inaccurate, and may itself be a symptom of nonadditivity or nonlinearity. Correlation between the independent variable and the error term generally indicates misspecification, whose effects may include bias, inefficiency, or inaccurate statistical inference.

In linear regression, the residuals are straightforwardly computed from the regression equation. In logistic regression, several different residuals are available, corresponding to the different levels (probability, odds, logit) at which the analysis may be conceptualized. The principal purpose for which residuals analysis is used in logistic regression is to identify cases for which the model works poorly, or cases that exert more than their share of influence on the estimated parameters of the model.

The difference between the observed and the predicted probability is $e_j = P(Y_j = 1) - \hat{P}(Y_j = 1)$, where $\hat{P}(Y_j = 1)$ is the estimated probability that $Y_j = 1$ based on the model. As Hosmer and Lemeshow explain, in linear regression, we can assume that the error is independent of the conditional mean of Y, but in logistic regression, the error variance is a function of the conditional mean. For this reason, residuals (estimates of error) are standardized by adjusting them for their standard errors. The Pearson (Hosmer & Lemeshow, 1989) or standardized (SPSS) or chi (SAS) residual is equal to

$$r_j = z_j = \chi_j = \frac{P(Y_j = 1) - \hat{P}(Y_j = 1)}{\sqrt{\hat{P}(Y_j = 1)[1 - \hat{P}(Y_j = 1)]}} \ .$$

This is just the difference between the observed and estimated probabilities divided by the binomial standard deviation of the estimated probability. For large samples, the standardized residual, hereafter z_j, should be normally distributed with a mean of 0 and a standard deviation of 1. Large positive or negative values of z_j indicate that the model fits a case j poorly. Because z_j should have a normal distribution, 95% of the cases should have values between -2 and $+2$, and 99% of cases should have values between -2.5 and $+2.5$.

An alternative or supplement to the Pearson residual is the deviance residual, which is equal to $d_j = -2\ln(\text{predicted probability of correct group})$. The deviance residual is the contribution of each case to the deviation statistic, D_M. Like z_j, d_j should have a normal distribution with a mean of 0 and a standard deviation of 1 for large samples. A third residual, the logit residual, is equal to the residual e_j divided by its variance (instead of its standard deviation, as in the standardized residual). This may be written

$$l_j = \frac{P(Y_j = 1) - \hat{P}(Y_j = 1)}{\hat{P}(Y_j = 1)[1 - \hat{P}(Y_j = 1)]} \ .$$

Nonnormality of Residuals

In OLS regression, it is usually assumed that the errors are normally distributed. In small samples, if this assumption is violated, it renders statistical inference based on the regression equation (e.g., the statistical significance of the regression coefficients) inaccurate. In large samples, inaccuracy of statistical inference is considered inconsequential because of results of the Central Limit Theorem, which, briefly, indicates that the distribution of the regression coefficients in repeated sampling for large enough samples will approach a normal distribution with known mean (equal to the population mean) and variance. In logistic regression, the errors are not assumed to have a normal distribution. Instead, it is assumed that the distribution of the errors follows a binomial distribution, which approximates a normal distribution only for large samples. If the residuals are used to estimate the errors, and if they are normally distributed (for a large sample), we can be more confident that our inferential statistics are correct, because normal (the distribution we are considering) and binomial (the assumed distribution) distributions are about the same for large samples. Contrary to the situation in linear regression analysis, however, if we find that the residuals are not normally distributed for small samples, we need not necessarily be concerned about the validity of our statistical inferences.

We can test for normality by plotting the standardized or deviance residuals against a normal curve, or in a normal probability plot; see, for example, Norusis (1990). More important, we can use the standardized and deviance residuals to identify cases for which the model fits poorly, cases with positive or negative standardized or deviance residuals greater than 2 in absolute value.[18] This may help us identify not only cases for which the model fits poorly but also cases that exert a disproportionately large influence on the estimates of the model parameters.

Detecting and Dealing With Influential Cases

Cases that have a large influence on the parameters of the logistic regression model may be identified by high values of the leverage statistic, or hat-value, h_i. In linear regression, the leverage statistic is derived from the equation $\hat{Y}_j = h_{1j}Y_1 + h_{2j}Y_2 + \ldots + h_{nj}Y_n = \sum_i h_{ij}Y_i$, and it expresses the predicted value of Y for a case j as a function of the observed values of Y for case j and for all of the other cases as well (Fox, 1991). Each coefficient h_{ij} captures the influence of the observed variable Y_i on the predicted value $\hat{Y}_j$. It can be shown that $h_{ii} = \sum (h_{ij})^2$, so if we designate $h_i = h_{ii}$, we have a measure of the overall influence of Y_i on the predicted values of Y for all of the cases in the sample. The leverage is similarly derived in logistic regression (Hosmer & Lemeshow, 1989, pp. 150-151), and it ranges from 0 (no influence) to 1 (it completely determines the parameters in the model). In an equation with k independent variables (including each design variable as a separate variable) or, equivalently, in an equation in which there are k degrees of freedom associated with G_M, the sum of the values of h_i is equal to $k + 1$, and the mean value of $h_i \sum h_i / N = (k + 1)/N$. Cases with hat values larger than $(k + 1)/N$ are influential cases.

Other indices of the influence of an individual case include the change in the Pearson χ^2 statistic and the change in D_M attributable to deleting the case from the analysis. The change in the Pearson χ^2 attributable to deleting a case j is $\Delta\chi_j^2 = z_j^2/(1 - h_j)$, where z_j is the standardized residual and h_j is the leverage statistic for case j. The change in D_M is equal to $\Delta D_j = d_j^2 - z_j^2 h_j/(1 - h_j) = d_j^2 - h_j(\Delta\chi_j^2)$, where d_j is the deviance residual. Both ΔD_j and $\Delta\chi_j^2$ have a chi-square distribution, and their values should be interpreted accordingly. Their respective square roots should have an approximately normal distribution; if $\sqrt{\Delta D_j}$ (the Studentized residual provided in SPSS LOGISTIC REGRESSION, or C in SAS PROC LOGISTIC) or $\sqrt{\Delta\chi_j^2}$ is less than −2 or greater than +2, it indicates a case that may be poorly fit and deserves closer inspection. The quantity $z_j^2 h_j/(1 - h_j)$ is itself an indicator of the overall change in regression estimates attributable to deleting an individual observation, and it is available in SAS PROC LOGISTIC as the optional statistic CBAR and in SPSS LOGISTIC REGRESSION as Cook's distance. A standardized version of this measure may be obtained by dividing Cook's distance by $(1 - h_j)$; $z_j^2 h_j/(1 - h_j)^2$ = dbeta, the standardized change in the regression coefficients attributable to the deletion of case j. The leverage statistic and the related statistics described above are all summary indicators of the influence of a case on the estimation of the model parameters. More detailed information can be obtained by examining changes in individual coefficients that occur when a case is deleted. The change in the logistic regression coefficient is described as the DFBETA in both SPSS and SAS.

Outliers and Residual Plots

Table 4.3 presents the results of an analysis of residuals. Cases with $\sqrt{\Delta D_j}$ less than −2 or greater than +2 were selected for examination. The table includes the sequential number of the case, the observed and predicted values of the case, the Pearson (ZResid), Studentized (SResid), and deviance (Dev) residuals, the leverage (Lever), and the deleted residuals ΔD_j (DIFDEV), $\Delta\chi^2$ (DIFCHI), and dbeta (DBETA). Part A of Table 4.3 presents the residuals for the model in Table 3.1. Part B presents the results with the most extreme outlier deleted. This case is one of the two identified in the analysis of nonlinearity in Figure 4.1. Part C presents the results with both of the outliers from Figure 4.1 deleted.

A first comment to be made about Table 4.3 is that several of the indicators are essentially redundant with one another. The change in the Pearson chi-square statistic, DIFCHI ($\Delta\chi^2$), is approximately equal to the Pearson residual, ZResid, squared. The deviance residual, Dev (d_j), is approximately equal to the Studentized residual (SResid), and the change in the deviance residual, DIFDEV (ΔD_j), is equal to the Studentized

TABLE 4.3
Logistic Regression Diagnostic Summaries

A. Full Model

CASE	Observed PMRJ5	Pred	ZResid	Dev	SResid	Lever	DIFCHI	DIFDEV	DBETA
66	1	.0991	3.0143	2.1500	2.1637	.0127	9.20	4.68	.12
94	0	.8608	-2.4864	-1.9858	-2.0325	.0455	6.48	4.13	.31
139	1	.0815	3.3565	2.2391	2.2668	.0243	11.55	5.14	.29
148	1	.0612	3.9183	2.3641	2.3762	.0102	15.51	5.65	.16
178	0	.9914	-10.7055	-3.0823	-3.0983	.0103	115.80	9.60	1.21
201	1	.0650	3.7937	2.3383	2.3526	.0121	14.57	5.53	.18

$G_M = 108.257$ 5 df $p = .0000$ $R^2_L = .367$

B. Most Extreme Case Deleted

CASE	Observed PMRJ5	Pred	ZResid	Dev	SResid	Lever	DIFCHI	DIFDEV	DBETA
1	0	.9122	-3.2239	-2.2059	-2.2557	.0436	10.87	5.09	.50
66	1	.0894	3.1913	2.1975	2.2116	.0127	10.32	4.89	.13
94	0	.8786	-2.6903	-2.0536	-2.0999	.0436	7.57	4.41	.34
139	1	.0861	3.2577	2.2146	2.2444	.0264	10.90	5.04	.30
148	1	.0510	4.3137	2.4396	2.4515	.0097	18.79	6.01	.18
200	1	.0661	3.7601	2.3312	2.3463	.0129	14.32	5.51	.19

$G_M = 118.156$ 5 df $p = .0000$ $R^2_L = .401$

C. Outliers From Figure 4.1 Deleted (Cases 178 and 1)

CASE	Observed PMRJ5	Pred	ZResid	Dev	SResid	Lever	DIFCHI	DIFDEV	DBETA
66	1	.0808	3.3732	2.2432	2.2573	.0125	11.52	5.10	.15
94	0	.8858	-2.7852	-2.0832	-2.1289	.0425	8.10	4.53	.36
133	0	.8675	-2.5584	-2.0105	-2.0506	.0387	6.81	4.20	.27
139	1	.0885	3.2099	2.2023	2.2332	.0275	10.59	4.99	.30
148	1	.0441	4.6536	2.4982	2.5097	.0092	21.86	6.30	.20
200	1	.0675	3.7175	2.3221	2.3378	.0135	14.01	5.47	.19

$G_M = 122.634$ 5 df $p = .0000$ $R^2_L = .416$

NOTE: Cases with studentized residuals greater than 2.0000000 are listed.

residual squared. The Pearson χ^2-based residuals are larger than the residuals based on the deviation χ^2, but they provide essentially the same information about the cases. The leverage and DBETA provide information not evident from the other diagnostics, similar to but not redundant with one another. Further analysis of the table focuses on the Pearson residual, the Studentized residual, the leverage, and DBETA.

In Part A, one case, number 178, stands out. The Pearson residual is an enormous -10.7, the Studentized residual is greater than 3 in absolute value, and DBETA is greater than 1, all indicators of an extremely poor fit. Deleting this case would result in an improvement in G_M of 9.899 (1 degree of freedom, $p = .003$) and an increase in R_L^2 of .034. Clearly the model would work better with this case deleted. In Part B, with case 178 deleted, no case stands out as clearly. Case 148 has the highest Pearson residual and the highest Studentized residual but would produce relatively little change in the logistic regression coefficients if deleted. Case 1 would have more of an effect on the logistic regression coefficients (DBETA = .50), and it has the fourth-highest Pearson residual and the third-highest Studentized and deviance residuals of the six cases selected as outliers. Case 1 has the additional feature that it is one of the two outliers identified in Figure 4.1 as introducing nonlinearity into the model. With case 1 deleted, G_M improves by 4.478 (1 degree of freedom, $p = .038$), and R_L^2 increases by .015. The improvement that results from deleting case 1 is considerably smaller than the improvement from removing case 178.

Should cases 1 and 178 be removed from the analysis? The answer to this question requires closer examination of the data. The two cases in question are both white, one male and one female. Both report low levels of belief that it is wrong to violate the law, but neither uses marijuana or hard drugs, and both report very low levels of alcohol use as well. Case 1 (female) has a slightly higher level of belief and a substantially lower level of exposure to delinquent friends than case 178 (male), and is therefore less inconsistent with the model than case 178. Although unusual, the results are plausible, and both cases should probably be retained. It would be useful, however, to extend the model to include variables that might explain why an individual who sees nothing wrong with breaking the law chooses not to use alcohol, marijuana, or other illicit drugs.

Landwehr, Pregibon, and Shoemaker (1984) and Hosmer and Lemeshow (1989) discuss graphical techniques for logistic regression diagnostics. These techniques offer a visual rather than numerical representation that may be more intuitively appealing to some researchers. For example, Hosmer and Lemeshow (1989) recommend plots of DIFCHI, DIFDEV, and

DBETA with the predicted values in order to detect outlying cases. Examples of these plots are provided in the first column of plots in Figure 4.3. Each of the three plots represents two curves, one declining from left to right (cases for which the observed value of PMRJ5 is 1) and one increasing from left to right (cases for which the observed value of PMRJ5 is 0). Cases in the upper left and right corners of the plot are cases for which the model fits poorly. In the plot of the χ^2 change (DIFCHI) with the predicted value, one case is an extreme outlier, with DIFCHI greater than 100. From Table 4.3, we can see that this is case 178. Similarly, case 178 is the case in the plot of DIFDEV with a value of DIFDEV greater than 8, and in the plot of DBETA with a value greater than 1. On the scale of the plots in the first column of Figure 4.3, all of the other cases seem to cluster fairly close together. Once case 178 is deleted, however, the scale of the plots may be changed, and other cases appear as outliers.

The second column of plots in Figure 4.3 presents the same plots with case 178 deleted. The plots may appear to be more spread out than the plots in the first column, but this is only because the scale has changed (from 0-100 to 0-20 for DIFCHI, from 0-10 to 0-6.25 for DIFDEV, and from 0-1.25 to 0-.5 for DBETA). The combination of the two curves for PMRJ5 = 1 and PMRJ5 = 0 takes on a characteristic goblet-shaped pattern (especially for DIFDEV), with the outliers again located in the upper right and left corners of the plot, and also in the "cup" of the goblet. In contrast to the first column, there is a relatively smooth transition from the upper corners to the rest of the graph, The most outlying cases are not as sharply separated as case 178 was in the first column. This is reflected numerically in Table 4.3, Part B, in which none of the Studentized residuals is larger than 2.5 in absolute value, and none of the DBETAs is greater than 1.

A Suggested Protocol for
Logistic Regression Diagnostics

Testing for collinearity should be a standard part of any logistic regression analysis. It is quick, is simple to implement with existing regression software, and may provide valuable information about potential problems in the logistic regression analysis before the analysis is undertaken. The Box-Tidwell test for nonlinearity is quick, easy to perform, and not overly sensitive to minor deviations from linearity, and it should also be incorporated as a standard procedure in logistic regression. Whether to test for nonadditivity by modeling interactions among the independent variables

78

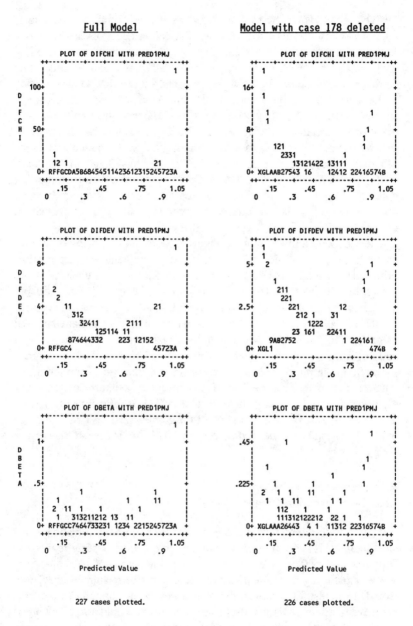

Figure 4.3. Logistic Regression Diagnostic Plots

depends on whether one has theoretical or other reasons to believe that such interactions exist. Modeling nonlinearity and nonadditivity should be approached with some caution, however. There is a real danger of overfitting a model, building in components that really capture random variation, rather than systematic regularities in behavior. Using logistic regression diagnostics, like using linear regression diagnostics, is more art than science. The diagnostic statistics hint at potential problems, but what those problems are, and whether remedial action is required, can only be decided after closer inspection of the data for the unusual cases. In a sample of 200-250, random sampling variation alone will produce 10-12 cases with values greater than 2 or less than −2 on standardized, normally distributed variables such as the deviance residual or the Studentized residual. Even cases with very large residuals, like case 178 in Table 4.3, do not necessarily indicate problems in the model, insofar as we are dealing with nondeterministic models in which individual human choice and free will may naturally produce less than perfect prediction of human behavior.

As a general approach, it seems appropriate to perform at least a limited set of diagnostics on any model, as a precaution against miscoded data and a guide to weaknesses in our conceptual models. A minimal set of diagnostics might include the Studentized residual, the leverage, and dbeta. Studentized residuals less than −3 and greater than +3 definitely deserve closer inspection; values less than −2 or greater than +2 may also warrant some concern. The disadvantage to the Pearson residual is that the information it provides tends to be redundant with the information provided by the deviance and Studentized residuals, and the deviance residual, not the Pearson residual, is the criterion for estimating the parameters of the model and is thus somewhat more pertinent to the analysis of residuals. The advantage to the Pearson residual is that, because it has larger values than the deviance residual, outlying cases sometimes stand out more sharply (as does case 178 in Part A of Table 4.3) with the Pearson residual than with the deviance or Studentized residual. Leverage values several times the expected value of $(k + 1)/N$ (which was about $5/227 = .02$ in this example) also deserve close attention. Large values of dbeta, especially values greater than 1 (remember, this is a standardized measure), also deserve closer examination. Whether the information contained in these diagnostics is presented visually is a matter of taste. The critical concern is that extreme values on these diagnostics require closer inspection of the data and possibly reconsideration of the model.

5. POLYTOMOUS LOGISTIC REGRESSION AND ALTERNATIVES TO LOGISTIC REGRESSION

Logistic regression analysis may be extended beyond the analysis of dichotomous variables to the analysis of categorical (nominal or ordinal) dependent variables with more than two categories. In the literature on logistic regression, the resulting models have been called polytomous, polychotomous, or multinomial logistic regression models. Here, the terms dichotomous and polytomous will be used to refer to logistic regression models, and the terms binomial and multinomial will be used to refer to logit models from which polytomous logistic regression models may be derived. For polytomous dependent variables, the logistic regression model may be calculated as a special case of the multinomial logit model (Agresti, 1990; Aldrich & Nelson, 1984; DeMaris, 1992; Knoke & Burke, 1980).

Mathematically, the extension of the dichotomous logistic regression model to polytomous dependent variables is straightforward. One value (typically the first or last) of the dependent variable is designated as the reference category, $Y = h_0$, and the probability of membership in other categories is compared with the probability of membership in the reference category. For nominal variables, this may be a direct comparison, like the indicator contrasts for independent variables in the logistic regression model for dichotomous variables. For an ordinal variable, contrasts may be made with successive categories, in a manner similar to repeated or Helmert contrasts for independent variables in dichotomous logistic regression models.

For dependent variables with some number of categories M, this requires the calculation of $M - 1$ equations, one for each category relative to the reference category, to describe the relationship between the dependent variable and the independent variables. For each category of the dependent variable except the reference category, we may write the equation

$$g_h(X_1, X_2, \ldots, X_k) = e^{a_h + b_{h1}X_1 + b_{h2}X_2 + \ldots + b_{hk}X_k}$$

$$h = 1, 2, \ldots, M - 1 \qquad (5.1)$$

where the subscript k refers, as usual, to specific independent variables X, and the subscript h refers to specific values of the dependent variable Y. For the reference category, $g_0(X_1, X_2, \ldots, X_k) = 1$. The probability that Y is equal to any value h other than the excluded value, h_0, is

$$P(Y = h \mid X_1, X_2, \ldots, X_k) = \frac{e^{a_h + b_{h1}X_1 + b_{h2}X_2 + \ldots + b_{hk}X_k}}{1 + \sum_{h=1}^{M-1} e^{a_h + b_{h1}X_1 + b_{h2}X_2 + \ldots + b_{hk}X_k}}$$

$$h = 1, 2, \ldots, M - 1 \qquad (5.2)$$

and for the excluded category, $h_0 = M$ or $h_0 = 0$,

$$P(Y = h_0 \mid X_1, X_2, \ldots, X_k) = \frac{1}{1 + \sum_{h=1}^{M-1} e^{a_h + b_{h1}X_1 + b_{h2}X_2 + \ldots + b_{hk}X_k}}$$

$$h = 1, 2, \ldots, M - 1. \qquad (5.3)$$

Note that when $M = 2$, we have the logistic regression model for the dichotomous dependent variable, the reference category is the first category, $h_0 = 0$, and we have a total of $M - 1 = 1$ equations to describe the relationship.

Neither SAS PROC LOGISTIC nor SPSS LOGISTIC REGRESSION is fully equipped to handle polytomous logistic regression models. It is possible, however, to use SAS PROC CATMOD and SPSS LOGLINEAR to calculate dichotomous logistic regression models, and to extend the techniques used in adapting CATMOD and LOGLINEAR for dichotomous logistic regression to the calculation of polytomous logistic regression models. Operationally, this involves calculating a multinomial logit model, then translating that model to obtain G_M, R_L^2, standardized coefficients, and prediction tables in the logistic regression format. For one special type of polytomous logistic regression model, a model with an ordinal dependent variable and equal slopes in each logistic function (each coefficient b is the same for $g_1, g_2, \ldots, g_{M-1}$), the model may be directly calculated using SAS PROC LOGISTIC. Operationally, the use of a polytomous dependent variable moves us further from OLS linear regression and closer to log-linear and logit models.

To illustrate the use of polytomous logistic regression, the dependent variable from previous examples, prevalence of marijuana use, is replaced by drug user type. Drug user type has four categories: Nonusers report that they have not used alcohol, marijuana, heroin, cocaine, amphetamines, barbiturates, or hallucinogens in the past year; alcohol users report having used alcohol, but no illicit drugs, in the past year; marijuana users report having used marijuana (and, except in one case, using alcohol as well); and polydrug users report illicit use of one or more of the "hard" drugs (heroin, cocaine, amphetamines, barbiturates, hallucinogens). Polydrug users also report using alcohol and, except in one case (a respondent who reported a single incident of hard drug use), marijuana as well.

The four categories can reasonably be regarded as being ordered from least serious to most serious drugs, in terms of legal consequences. Alternatively, with respect to the nonlegal consequences of the drugs, the scale could arguably be regarded as nominal. Both ordinal and nominal models of this variable will be considered. One additional change is made from previous models. Because the dependent variable has four categories, and because of the small number of cases in the category "other" on the variable ethnicity (ETHN), ethnicity was recoded into two categories, white and nonwhite, for the following analyses. Failure to do this would have resulted in problems with zero cells and with instability in estimates of coefficients and their standard errors.

Polytomous Nominal Dependent Variables

Table 5.1 presents the results of a logistic regression analysis with DRGTYPE as a dependent variable using a contrast for DRGTYPE that compares, in succession, (a) nonusers with alcohol users, (b) nonusers with marijuana users, and (c) nonusers with polydrug users. The resulting functions, $g_1(X)$, $g_2(X)$, and $g_3(X)$ may be defined as

$g_1 =$ logit (probability of using some alcohol versus nonuse of drugs),

$g_2 =$ logit (probability of using marijuana versus nonuse of drugs), and

$g_3 =$ logit (probability of using other illicit drugs versus nonuse of drugs).

The equations for g_1, g_2, and g_3 using unstandardized coefficients are, from Table 5.1,

TABLE 5.1
Polytomous Logistic Regression: Nominal Dependent Variable

Dependent Variable	Model Fit	Independent Variable	Unstandardized Coefficient	Statistical Significance of Unstandardized Coefficient	Standardized Coefficient
DRGTYPE	$G_M = 169.348$ $(p < .001)$ $R_L^2 = .282$				
(1) Alcohol	$R_1^2 = .189$	EDF5	.165	.112	.209
user vs.		BELIEF4	−.271	.000	−.319
nonuser		SEX (M)	.505	.194	.075
		ETHN			
		WHITE	1.616	.000	.202
		Intercept	6.146	.030	———
(2) Marijuana	$R_2^2 = .149$	EDF5	.506	.000	.671
user vs.		BELIEF4	−.285	.001	−.350
nonuser		SEX (M)	−.920	.067	−.143
		ETHN			
		WHITE	.357	.503	.047
		Intercept	2.221	.471	———
(3) Polydrug	$R_3^2 = .337$	EDF5	.633	.000	.677
user vs.		BELIEF4	−.360	.000	−.357
nonuser		SEX (M)	−2.224	.001	−.279
		ETHN			
		WHITE	2.209	.000	.233
		Intercept	.761	.826	———
	$R_0^2 = .303$				

$$g_1 = .165(\text{EDF5}) - .271(\text{BELIEF4}) + .505(\text{SEX}) + 1.616(\text{WHITE}) + 6.146,$$

$$g_2 = .506(\text{EDF5}) - .285(\text{BELIEF4}) - .920(\text{SEX}) + .357(\text{WHITE}) + 2.221, \quad\text{and}$$

$$g_3 = .633(\text{EDF5}) - .360(\text{BELIEF4}) - 2.224(\text{SEX})$$
$$+ 2.209(\text{WHITE}) + .761.$$

Neither D_0 nor G_M is directly available from SAS PROC CATMOD or SPSS LOGLINEAR. For both, the likelihood chi-square statistics are based on comparisons of cells in a contingency table, rather than on probabilities of category membership. The two are related, however, and it is possible to derive the statistics appropriate for logistic regression analysis from the statistics provided by the log-linear analysis programs. For SAS PROC CATMOD, the appropriate steps are:

1. Compute $D_0 = \sum(n_{Y=h})\ln[\text{P}(Y=h)] = \sum(n_{Y=h})\ln(n_{Y=h}/N)$, where $n_{Y=h}$ is the number of cases for which Y is equal to one of its possible values, h; N is the total sample size; and the sum is taken over all possible values, h, of Y.

2. Examine the iteration history of the model; the "−2 log-likelihood" from the final iteration, under the heading "MAXIMUM LIKELIHOOD ANALYSIS," is approximately (but not exactly) equal to the deviation χ^2, D_M.

3. Compute $G_M = D_0 - D_M$; compute $R_L^2 = G_M/D_0$.

If these procedures are followed with a dichotomous dependent variable, the resulting figures are approximately equal to the G_M and R_L^2 that would be obtained in the identical analysis from SAS PROC LOGISTIC.[19] An alternative method of estimating G_M is to calculate two models, one with only the intercept and a second with all of the independent variables included (both on the same set of cases), and calculate the difference between the "−2 log-likelihood" from the final iteration for each model. This is the procedure used for calculating G_M and R_L^2 in SPSS:

1. Compute $D_0 = \sum(n_{Y=h})\ln[\text{P}(Y=\text{h})] = \sum(n_{Y=h})\ln(n_{Y=h}/N)$, where $n_{Y=h}$ is the number of cases for which Y is equal to one of its possible values, h; N is the total sample size; and the sum is taken over all possible values, h, of Y.

2. Estimate two models, the first one with only the intercept (e.g., /DESIGN = DRGTYPE/) and the second one including the predictors (/DESIGN = DRGTYPE DRGTYPE BY EDF5 DRGTYPE BY BELIEF4 DRGTYPE BY SEX DRGTYPE BY ETHN/).

3. Subtract the likelihood ratio chi-square for the second model from the likelihood ratio chi-square for the first: G_M = Intercept likelihood ratio χ^2 − Model likelihood ratio χ^2. Compute $R_L^2 = G_M/D_0$.

In SPSS, for a dichotomous dependent variable, the results from this procedure will be exactly equal to the G_M and R_L^2 that would be obtained for the same model if it could be run in SPSS LOGISTIC REGRESSION.[20] The process is a bit awkward but, if the likelihood ratio statistics provided in CATMOD and LOGLINEAR are used without modification, they produce results different from those that would be obtained using SPSS LOGISTIC REGRESSION or SAS PROC LOGISTIC when the dependent variable is dichotomous.

The calculation of R^2 or η^2 and the standardized logistic regression coefficients is done separately for each logistic function, g_1, g_2, and g_3. (This is similar to the calculation of separate canonical correlation coefficients and standardized discriminant function coefficients for each linear discriminant function in discriminant analysis; see Klecka, 1980.) Prediction tables can be constructed by calculating the probability of classification for each value of Y, including the reference category, using Equations 5.2 and 5.3, then classifying each case into the category of Y for which it has the highest probability. The table itself can then be constructed using any contingency table routine such as SAS PROC FREQ or SPSS CROSSTABS. Part A of Figure 5.1 shows the classification table associated with Table 5.1. Once the classification table has been constructed, indices of predictive efficiency can be calculated, as they have been for Figure 5.1, using the procedures described in Chapter 2. It is for polytomous models with nominal dependent variables that the differences between λ_p and τ_p, as opposed to other proposed indices of predictive efficiency, become most evident.

For Table 5.1, the model fits fairly well, as indicated by the statistically significant model χ^2 and the R_L^2 of .28. The explained variance in logit(Y) varies by the category of the dependent variable and is highest for g_3 (polydrug use) and lowest for g_2 (marijuana use). For alcohol use, the standardized coefficients indicate that the best predictor is belief that it is wrong to violate the law, followed by ethnicity (white respondents are more likely to use alcohol than nonwhites). Neither exposure to delinquent friends nor gender has a statistically significant effect on the distinction between nonusers and alcohol users. For both marijuana and polydrug use, the best predictor is exposure to delinquent friends, followed by belief, then gender. Ethnicity is not a statistically significant predictor for marijuana use, but white respondents are more likely than nonwhite respondents to be polydrug users. As Figure 5.1 indicates, the indices of predictive efficiency λ_p and τ_p are both statistically significant and moderately strong.

A: Classification Table for Table 5.1, Polytomous Nominal Logistic Regression

		Predicted				Percentage correct
		1	2	3	4	
	1	17	41	1	0	28.8
	2	2	83	1	1	95.4
Observed	3	1	35	7	7	14.0
	4	0	14	0	17	54.8
Total						54.6

$$\lambda_p = .264 \quad d = 5.051 \quad p = .000$$
$$\tau_p = .368 \quad d = 8.865 \quad p = .000$$

B: Classification Table for Figure 5.2, Polytomous Ordinal Logistic Regression

		Predicted				Percentage correct
		1	2	3	4	
	1	23	36	0	0	39.0
	2	11	76	0	0	87.4
Observed	3	1	49	0	0	0.0
	4	0	31	0	0	0.0
Total						43.6

$$\lambda_p = .086 \quad d = 1.638 \quad p = .051$$
$$\tau_p = .215 \quad d = 5.176 \quad p = .000$$

Figure 5.1. Classification Tables for Polytomous Logistic Regression

Polytomous or Multinomial Ordinal Dependent Variables

When the dependent variable is measured on an ordinal scale, many possibilities for analysis exist, including but by no means limited to logistic regression analysis. For a more complete discussion, see Agresti (1984) and Clogg and Shihadeh (1994). Briefly, the options available include

1. ignoring the ordering of the categories of the dependent variable and treating it like a nominal variable,
2. treating the variable as though it were measured on a true ordinal scale,
3. treating the variable as though it were measured on an ordinal scale, but the ordinal scale represented crude measurement of an underlying interval/ratio scale, and
4. treating the variable as though it were measured on an interval scale.

One possibility consistent with the first option is the use of a multinomial logit or logistic regression model for a nominal categorical dependent variable, as in Table 5.1. Also possible under option 1 would be the use of discriminant analysis (Klecka, 1980). An example of the second option would be the use of a *cumulative logit* model, in which the transformation of the dependent variable incorporated not only each category compared with a reference category but also a comparison of each category with all of the categories with higher (or lower) numeric codes than the present category. The third option, assuming an underlying interval scale, could be implemented in LISREL by using weighted least squares (WLS) analysis of polychoric correlations (Jöreskog & Sörbom, 1988). The fourth option might be implemented by using OLS regression with an ordinal dependent variable.

Selecting one of the options is a matter requiring careful judgment. The fourth option effectively assumes that the data are measured more precisely than they really are, but for ordinal variables with a large number of categories, it may be reasonable. The use of WLS with polychoric correlations appears to be a better option; it can be used with both large and small numbers of categories, and for most ordinal variables. The assumption of imprecise measurement of a quantity that is really continuous (political conservatism, seriousness of drug use) is inherently plausible. Both of these options allow predicted values that lie outside the range of observed values but, under the assumption of imprecise measurement, this may be reasonable.

Mechanical application of options available in existing software packages is not recommended. For example, SAS PROC LOGISTIC can calculate polytomous logistic regression models for ordinal dependent variables, but PROC LOGISTIC assumes a *parallel slopes model* for the dependent variable. This model assumes that the coefficient for each independent variable is invariant across the three equations, that is, $b_{EDF5,1} = b_{EDF5,2} = b_{EDF5,3}$, $b_{SEX,1} = b_{SEX,2} = b_{SEX,3}$, where the variable in the subscript is the variable to which the coefficient refers and the number in the subscript is the equation (1, 2, or 3) in which the coefficient appears. For the parallel slopes model, only the intercept is different for the three equations; otherwise, the effects of the independent variables are assumed to be constant across group comparisons. *It is important to emphasize that although this model is easily calculated using SAS PROC LOGISTIC, it may not be the most appropriate model for the relationship between the dependent variable and the predictors.*

Figure 5.2 summarizes the results of analyzing drug user type, DRGTYPE, as an ordinal variable in SAS PROC LOGISTIC. SAS provides a test of the assumption that the slopes are equal, the Score test. For Figure 5.2 the Score test of the null hypothesis that the slopes are equal is 32.066 with 8 degrees of freedom, statistically significant at the .0001 level. Because the Score test is statistically significant, the parallel slopes assumption is rejected, indicating that a model that does not assume parallel slopes would be more appropriate. The reasons for the rejection of the equal slopes model are evident from Table 5.1: the variation in both the strength and statistical significance of the effects of EDF5 (not statistically significant for alcohol users as opposed to nonusers), SEX (not statistically significant for alcohol users as opposed to nonusers; stronger for polydrug users than for marijuana users as opposed to nonusers), and ETHN (not statistically significant for marijuana users as opposed to nonusers). The pattern of the differences in the coefficients in Table 5.1 (especially the down-and-up pattern of the coefficients for ethnicity) suggests that treating DRGTYPE as a categorical nominal variable may be the best option.

Conclusion: Polytomous Dependent Variables and Logistic Regression

The principal concern in using logistic regression analysis with polytomous dependent variables is not how to make the model work but instead whether the logistic regression model is appropriate at all. For ordinal dependent variables, the problems that motivated the development of the logistic regression model (out-of-range predicted values of the dependent variable, heteroscedasticity) may not be present, and other models may be more appropriate than logistic regression, depending on assumptions about the underlying scale of the dependent variable and the functional form (linear, monotonic, nonmonotonic) of the relationship between the depen-

```
data;
   infile 'saslr16a.dat' missover linesize=60 firstobs=1 obs=257;
   input ID F66 SEX 8 ETHN 10 USR5 12 PDRUGS5 14-15 PMRJ5 17-18 PALC5 20-21
         DRGTYP5 23-24 EDP5 26-33 BELIEF4 35-42 MEANSCIN 44-51 MEANFAIN 53-60;
   if ethn=1 then white=1; if ethn=2 then white=0; if ethn=3 then white=0;
   if drgtyp5=1 then drgtyp5r=4; if drgtyp5=2 then drgtyp5r=3;
   if drgtyp5=3 then drgtyp5r=2; if drgtyp5=4 then drgtyp5r=1;const5=1;
run;
proc logistic;
   model drgtyp5r=edp5 belief4 sex white;run;
```

Data Set: WORK.DATA1
Response Variable: DRGTYP5R
Response Levels: 4
Number of Observations: 227
Link Function: Logit

Response Profile

Ordered Value	DRGTYP5R	Count
1	1	31
2	2	50
3	3	87
4	4	59

WARNING: 30 observation(s) were deleted due to missing values for the response or explanatory variables.

Score Test for the Proportional Odds Assumption
Chi-Square = 32.0660 with 8 DF (p=0.0001)

Criteria for Assessing Model Fit

Criterion	Intercept Only	Intercept and Covariates	Chi-Square for Covariates	
				R^2_1 = .004
				R^2_2 = .089
				R^2_3 = .151
AIC	606.600	485.626	.	R^2_0 = .264
SC	616.875	509.600	.	
-2 LOG L	600.600	471.626	128.975 with 4 DF (p=0.0001)	R_L^2 = .215
Score	.	.	97.395 with 4 DF (p=0.0001)	

Analysis of Maximum Likelihood Estimates

Variable	DF	Parameter Estimate	Standard Error	Wald Chi-Square	Pr > Chi-Square	Standardized Estimate b_{ik}	Odds Ratio	Standardized coefficient $b^*=(b)(s_x)/s_t$
INTERCP1	1	-1.3616	1.4611	0.8684	0.3514	.	0.256	.
INTERCP2	1	0.6157	1.4513	0.1800	0.6714	.	1.851	.
INTERCP3	1	2.9884	1.4655	4.1583	0.0414	.	19.854	.
EDP5	1	0.2701	0.0424	40.5402	0.0001	0.633781	1.310	.343
BELIEF4	1	-0.1774	0.0426	17.3225	0.0001	-0.386429	0.837	-.209
SEX	1	-0.7905	0.2630	9.0312	0.0027	-0.218288	0.454	-.118
WHITE	1	0.8343	0.3167	6.9391	0.0084	0.193729	2.303	.105

Association of Predicted Probabilities and Observed Responses

Concordant = 80.5%	Somers' D	= 0.623
Discordant = 18.2%	Gamma	= 0.631
Tied = 1.3%	Tau-a	= 0.449
(18509 pairs)	c	= 0.81

Figure 5.2. SAS Output for Ordinal Logistic Regression

dent variable and the independent variables. If there is an underlying interval scale, and if the relationships appear to be linear or monotonic, weighted least squares with polychoric correlations may be the best option.

For nonmonotonic relationships, and especially when there are relatively few categories of the dependent variable, it may be best to treat the dependent variable as though it were nominal.

When the dependent variable is nominal, or is an ordinal variable with few categories and is treated as though it were nominal, the logistic regression model may be calculated as a multinomial logit model in statistical programs such as SAS PROC CATMOD and SPSS LOGLINEAR. These programs allow the researcher to distinguish between categorical ordinal or nominal variables and continuous interval or ratio variables (using the "DIRECT" command in SAS PROC CATMOD, or defining the variable as having a single degree of freedom in SPSS LOGLINEAR), and it is possible to use different contrasts to examine different patterns of effects or comparisons. If this is the option selected, it is helpful to have some knowledge of log-linear and logit models (Aldrich & Nelson, 1984; DeMaris, 1992; Knoke & Burke, 1980) and their implementation in SAS or SPSS. An alternative worth considering is discriminant analysis (Klecka, 1980). Another alternative, separate logistic regressions (Begg & Grey, 1984; Hosmer & Lemeshow, 1989, pp. 230-232), does not appear to produce results sufficiently consistent with the multinomial logit/polytomous logistic regression model to warrant its use except for exploratory or diagnostic purposes. Separate logistic regressions may, however, be a useful supplement to polytomous logistic regression to provide more detail about the pattern of relationships in the model.

Generally, the extension of the logistic regression model to the analysis of polytomous nominal dependent variables is straightforward, requiring only some computation based on information provided by log-linear analysis programs (CATMOD and LOGLINEAR) to obtain the statistics relevant to logistic regression models, as opposed to multinomial logit models. Alternatively, the use of polytomous logistic regression models may mark the point at which familiarity with old standbys like SAS and SPSS is outweighed by the greater ease of use of newer or more specialized statistical packages such as STATA (Hamilton, 1993), which allows direct calculation of polytomous logistic regression models for nominal (MLOGIT) or ordinal (OLOGIT) dependent variables.

Conclusion

The power and current popularity of logistic regression analysis make it particularly susceptible to misuse. Thoughtless and mechanical applications of logistic regression analysis will be no more fruitful than thoughtless and mechanical applications of linear regression or any other technique. It is important to recognize the weaknesses as well as the strengths of the method. Logistic regression is especially appropriate for the analysis of dichotomous and unordered nominal polytomous dependent variables. For dichotomous variables, logistic regression models can easily be calculated using existing, widely available software, but as this is being written, some of the statistics that would be useful in a logistic regression analysis (standardized logistic regression coefficients, R_L^2, η^2 or R^2, and indices of predictive efficiency) must be computed by hand. For polytomous nominal dependent variables, it is necessary in SAS and SPSS to use programs for log-linear analysis to calculate logistic regression coefficients, and to use information from those programs to calculate R_L^2, R^2 or η^2, indices of predictive efficiency, and standardized logistic regression coefficients by hand. For ordinal polytomous dependent variables, it may be possible to use polytomous logistic regression analysis, but other models including linear regression and weighted least squares with polychoric correlations also deserve serious consideration. Polytomous ordinal variables are the dependent variables for which the technical motivation for using logistic regression is weakest and for which alternative methods of analysis are most likely to provide better solutions than logistic regression.

The necessity of supplementing or working around existing logistic regression software in order to obtain many of the statistics that are useful in evaluating the logistic regression model is, one can hope, a temporary situation reflecting the relative novelty of logistic regression software in widely available, general purpose statistical packages. SAS PROC LOGISTIC was new to version 6 (1989) of SAS and replaced (with changes) an older, user-contributed procedure available through the SAS supplemental library, PROC LOGIST (Harrell, 1986). SPSS LOGISTIC REGRESSION became available with version 4 (1990) of SPSS (previously SPSS-X). One need only examine the social science literature to realize how rare it is to find any reference to the use of logistic regression analysis prior to 1990. Even the technical literature on logistic regression makes little reference to some of the topics (standardized coefficients, indices of predictive efficiency, the use of η^2 or R^2 with logistic regression) covered in this monograph. One can hope that many of the "kludges" for making logistic regression analysis work with existing software detailed in this monograph (especially the calculation of G_M and R_L^2 for polytomous models) will become obsolete as the software available for logistic regression analysis is expanded and improved.

NOTES

1. Although the relationship being modeled often represents a causal relationship, in which the single predicted variable is believed to be an effect of the one or more predictor variables, this is not always the case. One can as easily predict a cause from an effect (e.g., predict whether different individuals are male or female based on their income) as predict an effect from a cause (predicting income based on whether someone is male or female). Throughout this monograph, the emphasis is on predictive rather than causal relationships, although the language of causal relationships is sometimes employed. Describing a variable as independent or dependent, therefore, or as an outcome or a predictor, does not necessarily imply a causal relationship. Instead, all relationships should be regarded as definitely predictive but only possibly causal in nature.

2. Data are taken from the National Youth Survey, a national household probability sample of individuals who were adolescents, age 11-17, in 1976, and young adults, age 21-26, in 1986. Data were collected annually for the years 1976 to 1980, then in 3-year intervals thereafter, including 1983 and 1986. The data include information on self-reported illegal behavior, family relationships, school performance, and sociodemographic characteristics of the respondents. Details on sampling and measurement may be found in Elliott et al. (Elliott, Huizinga, & Ageton, 1985; Elliott, Huizinga, & Menard, 1989). For present purposes, attention is restricted to respondents who were 16 years old in 1980. In the scatterplot, the numbers and symbols refer to numbers of cases at a given point on the plot. A 1 indicates one case, a 2 indicates two cases, a 9 indicates nine cases. The letters A-Z continue the count: A = 10 cases, B = 11 cases, . . . , Z = 35 cases. When more than 35 cases occupy a single point, an asterisk (*) is used. The plot in this figure and plots in other figures in this monograph were generated by the SPSS PLOT procedure.

3. For a review of levels of measurement, see, for example, Agresti and Finlay (1986, pp. 14-18).

4. The *unconditional mean* of Y is simply the familiar mean, $\bar{Y} = \sum Y_j / N$. The *conditional mean* of Y for a given value of X is calculated by selecting only those cases for which X has a certain value and calculating the mean for those cases. The conditional mean can be denoted $\bar{Y}_{X=i} = \sum Y_{ij} / n_i$, where i is the value of X for which we are calculating the conditional mean of Y, Y_{ij} are the values of Y for the cases ($j = 1, 2, \ldots, n_i$) for which $X = i$, and n_i is the number of cases for which $X = i$.

5. A brief discussion of probability, including conditional probabilities, is presented in Appendix A.

6. The logarithmic transformation is one of several possibilities discussed by Berry and Feldman (1985, pp. 63-64), Lewis-Beck (1980, p. 44), and others to deal with relationships that are nonlinear with respect to the variables but may be expressed as linear relationships with respect to the parameters.

7. SPSS LOGISTIC REGRESSION and SAS PROC PROBIT use variants of the Newton-Raphson algorithm, and SAS PROC LOGISTIC uses an iteratively reweighted least-squares algorithm. Introductory treatment of these methods and references to more technical coverage may be found in Agresti (1990) or Eliason (1993).

8. The designation ϕ_p was selected because ϕ_p, like ϕ, is based on comparisons between observed and expected values for individual cells (rather than rows or columns, as with λ_p and τ_p), because the numerical value of ϕ_p is close to the numerical value of ϕ for tables with consistent marginals (row and column totals in which the larger row total corresponds to the larger column total) and ϕ_p and ϕ have the same sign (because they have the same numerator).

9. Phi-p can be adjusted by adding the minimum number of errors, $|(a + b) - (a + c)| = |b - c|$, to the expected number of errors without the model. This results in a coefficient that (a) retains the proportional change in error interpretation (because the adjustment is built into the calculation of the expected error) and (b) still may have negative values if the model is pathologically inaccurate. For extremely poor models, the revised index still has maximum values less than 1, even when the maximum number of cases is correctly classified and the increment over ϕ_p is small. Based again on similarities with ϕ, we may designate this adjusted ϕ_p as ϕ'_p. Note, however, that ϕ'_p cannot be calculated as $\phi_p / \max(\phi_p)$; to do so would destroy the proportional change in error interpretation for the measure and would leave the measure undefined when the maximum value of ϕ_p was zero.

10. For a two-tailed test, the null hypothesis is that there is no difference between the proportion of errors *with* the prediction model and the proportion of errors *without* the prediction model, and the alternative hypothesis is that the proportion of errors *with* the prediction model is not equal to the proportion of errors *without* the prediction model. For a one-tailed test, specifying that the model results in increased accuracy of prediction of the dependent variable, the null hypothesis is that the proportion of errors *with* the prediction model is no smaller than the proportion of errors *without* the prediction model, and the alternative hypothesis is that the proportion of errors *with* the prediction model is less than the proportion of errors *without* the prediction model. If we want to know whether the prediction model improves our ability to predict the classification of the cases, the one-tailed test is more appropriate, and a negative value of λ_p will result in a negative value for d and failure to reject the null hypothesis.

11. Copas and Loeber (1990) noted this property and indicated that in this situation it would be a misinterpretation to regard a value of 1 as indicating perfect prediction. This leads to two questions. How should we interpret the value of RIOC in this situation? What value does indicate perfect prediction for RIOC? Ambiguity of interpretation is an undesirable quality in any measure of change, and there are enough better alternatives that the use of the RIOC measure should be avoided.

12. It will not always be the case that logistic regression produces a higher R^2 than linear regression for a dichotomous dependent variable. In a parallel analysis of theft for the full National Youth Survey sample, R^2 for linear and logistic regression were .255 and .253, respectively.

13. This is sometimes called a "Type II" error or a false negative (failure to detect a relationship that exists), as opposed to a "Type I" error or a false positive (concluding that there is a relationship when there really is none).

14. This was done using the backward stepwise procedure, to be discussed later in the text.

15. If they were, they would indicate that non-Hispanic European Americans have the lowest rates of marijuana use, followed by African Americans, and Others have the highest prevalence of marijuana use. It is always questionable, however, to make statements about the nature of a relationship that is not statistically significant and may reflect nothing more than random sampling error.

16. Mathematically, the omitted category is redundant, of little or no interest. In both theory testing and applied research, however, it makes more sense to provide full information about the coefficients and their statistical significance for all three categories, rather than leave one for pencil and paper calculation.

17. Hosmer and Lemeshow (1989) distinguish between analyzing residuals based on individuals and analyzing residuals based on covariate patterns, the combinations of values of the independent variables that actually occur in the sample. When the number of covariate patterns is equal to the number of cases, or very nearly so, residuals must be analyzed for each case separately. This is the implicit approach taken in this section and in SAS PROC LOGISTIC and SPSS LOGISTIC REGRESSION. When the number of cases is much larger than the number of covariate patterns, or when some of the covariate patterns hold for more than five cases, Hosmer and Lemeshow recommend aggregating the cases by covariate pattern because of potential underestimation of the leverage statistic, h_j.

18. In a standard normal distribution with a mean of 0 and a standard deviation of 1, 95% of the cases should have standardized scores (or, in this context, standardized residuals) between -2 and $+2$, and 99% should have scores or residuals between -2.5 and $+2.5$. Having a standardized or deviance residual larger than 2 or 3 does not necessarily mean that there is something wrong with the model. We would expect about 5% of the sample to lie outside the range -2 to $+2$, and 1% to lie outside the range -2.5 to $+2.5$. Values far outside this range, however, are usually indications that the model fits poorly for a particular case, and they suggest either that there is something unusual about the case that merits further investigation or that the model may need to be modified to account for whatever it is that explains the poor fit for some of the cases.

19. SAS PROC LOGISTIC uses an iteratively reweighted least squares algorithm to calculate the model parameters; PROC CATMOD uses weighted least squares or maximum likelihood estimation, depending on the type of model being calculated.

20. SPSS LOGISTIC REGRESSION and SPSS LOGLINEAR both use a Newton-Raphson maximum likelihood estimation technique to estimate the model parameters.

APPENDIX A: PROBABILITIES

The probability of an event is estimated by its relative frequency in a population or sample. For example, if $n_{Y=1}$ is the number of cases for which $Y = 1$ in a sample and N is the total number of cases in the sample, then

1. We denote the probability that Y is equal to 1 as $P(Y = 1)$
2. $P(Y = 1) = n_{Y=1}/N$
3. The probability that Y is not equal to 1 is $P(Y \neq 1) = 1 - P(Y = 1) = 1 - (n_{Y=1}/N) = (N - n_{Y=1})/N$
4. The minimum possible value for a probability is 0 ($n_{Y=1} = 0$ implies $n_{Y=1}/N = 0$)
5. The maximum possible value for a probability is 1 ($n_{Y=1} = N$ implies $n_{Y=1}/N = 1$).

The *joint probability* of two independent events (occurrences that are unrelated to one another) is the product of their individual probabilities. For example, the probability that both X and Y are equal to 1, if X and Y are unrelated, is $P(Y = 1$ and $X = 1) = P(Y = 1) \times P(X = 1)$. If X and Y are related (e.g., if the probability that Y is equal to 1 depends on the value of X), then $P(Y = 1$ and $X = 1)$ will not be equal to $P(Y = 1) \times P(X = 1)$. Instead, we will want to consider the *conditional probability* that $Y = 1$ when $X = 1$, or $P(Y = 1 \mid X = 1)$.

The conditional probability that $Y = 1$ is the probability that $Y = 1$ *for a given value of some other variable*. [In this context, we may sometimes refer to $P(Y = 1)$, the probability that $Y = 1$ regardless of the value of any other variable, as the *unconditional* probability that $Y = 1$.] For example, the probability that the prevalence of marijuana use is equal to 1 for the data in Figure 2.1 is $P(PMRJ5 = 1) = .35$ (for males and females combined; detailed data not shown). The conditional probability that prevalence of marijuana use is equal to 1 is $P(PMRJ5 = 1 \mid SEX = 0) = .45$ for females and $P(PMRJ5 = 1 \mid SEX = 1) = .25$ for males. For a dichotomous variable, coded as 0 or 1, the probability that the variable is equal to 1 is equal to the mean for that variable, and the conditional probability that the variable is equal to 1 is equal to the conditional mean (see note 4) for the variable.

REFERENCES

AGRESTI, A. (1984) *Analysis of Ordinal Categorical Data.* New York: Wiley.

AGRESTI, A. (1990) *Categorical Data Analysis.* New York: Wiley.

AGRESTI, A., and FINLAY, B. (1986) *Statistical Methods for the Social Sciences* (2nd edition). San Francisco: Dellen.

ALDRICH, J. H., and NELSON, F. D. (1984) *Linear Probability, Logit, and Probit Models.* Sage University Paper series on Quantitative Applications in the Social Sciences, 07-045. Beverly Hills, CA: Sage.

BEGG, C. B., and GREY, R. (1984) "Calculation of polychotomous logistic regression parameters using individualized regressions." *Biometrika* 71: 11-18.

BENDEL, R. B., and A. A. AFIFI (1977) "Comparison of stopping rules in forward regression." *Journal of the American Statistical Association* 72: 46-53.

BERRY, W. D. (1993) *Understanding Regression Assumptions.* Sage University Paper series on Quantitative Applications in the Social Sciences, 07-092. Newbury Park, CA: Sage.

BERRY, W. D., and FELDMAN, S. (1985) *Multiple Regression in Practice.* Sage University Paper series on Quantitative Applications in the Social Sciences, 07-050. Beverly Hills, CA: Sage.

BOHRNSTEDT, G. W., and KNOKE, D. (1988) *Statistics for Social Data Analysis.* Itasca, IL: F. E. Peacock.

BOLLEN, K. A. (1989) *Structural Equation Models With Latent Variables.* New York: Wiley.

BULMER, M. G. (1979) *Principles of Statistics.* New York: Dover.

CLOGG, C. C., and SHIHADEH, E. S. (1994) *Statistical Models for Ordinal Variables.* Thousand Oaks, CA: Sage.

COPAS, J. B., and LOEBER, R. (1990) "Relative improvement over chance (RIOC) for 2×2 tables." *British Journal of Mathematical and Statistical Psychology* 43: 293-307.

COSTNER, H. L. (1965) "Criteria for measures of association." *American Sociological Review* 30: 341-353.

DeMARIS, A. (1992) *Logit Modeling.* Sage University Paper series on Quantitative Applications in the Social Sciences, 07-086. Newbury Park, CA: Sage.

ELIASON, S. R. (1993) *Maximum Likelihood Estimation: Logic and Practice.* Sage University Paper series on Quantitative Applications in the Social Sciences, 07-096. Newbury Park, CA: Sage.

ELLIOTT, D. S, HUIZINGA, D., and AGETON, S. S. (1985) *Explaining Delinquency and Drug Use.* Beverly Hills, CA: Sage.

ELLIOTT, D. S., HUIZINGA, D., and MENARD, S. (1989) *Multiple Problem Youth.* New York: Springer-Verlag.

FARRINGTON, D. P., and LOEBER, R. (1989) "Relative Improvement Over Chance (RIOC) and Phi as measures of predictive efficiency and strength of association in 2 $\times$ 2 tables." *Journal of Quantitative Criminology* 5: 201-213.

FOX, J. (1991) *Regression Diagnostics.* Sage University Paper series on Quantitative Applications in the Social Sciences, 07-079. Newbury Park, CA: Sage.

HAGLE, T. M., and MITCHELL, G. E., II (1992) "Goodness-of-fit measures for probit and logit." *American Journal of Political Science* 36: 762-784.

HAMILTON, L. C. (1993) *Statistics With STATA 3*. Belmont, CA: Duxbury.

HARDY, M. (1993) *Regression With Dummy Variables*. Sage University Paper series on Quantitative Applications in the Social Sciences, 07-093. Newbury Park, CA: Sage.

HARRELL, F. E., Jr. (1986) "The LOGIST procedure." In SAS Institute, Inc. (Ed.), *SUGI Supplemental Library User's Guide* (Version 5 edition, pp. 269-293). Cary, NC: SAS Institute, Inc.

HOSMER, D. W., and LEMESHOW, S. (1989) *Applied Logistic Regression*. New York: Wiley.

JÖRESKOG, K. G., and SÖRBOM, D. (1988) *PRELIS: A Program for Multivariate Data Screening and Data Summarization* (2nd edition). Chicago: Scientific Software.

JÖRESKOG, K. G., and SÖRBOM, D. (1993) *LISREL 8: Structural Equation Modeling With the SIMPLIS Command Language*. Chicago: Scientific Software.

KLECKA, W. R. (1980) *Discriminant Analysis*. Sage University Paper series on Quantitative Applications in the Social Sciences, 07-019. Beverly Hills, CA: Sage.

KNOKE, D., and BURKE, P. J. (1980) *Log-Linear Models*. Sage University Paper series on Quantitative Applications in the Social Sciences, 07-020. Beverly Hills, CA: Sage.

LANDWEHR, J. M., PREGIBON, D., and SHOEMAKER, A. C. (1984) "Graphical methods for assessing logistic regression models." *Journal of the American Statistical Association* 79: 61-71.

LEWIS-BECK, M. S. (1980) *Applied Regression: An Introduction*. Sage University Paper series on Quantitative Applications in the Social Sciences, 07-022. Beverly Hills, CA: Sage.

LOEBER, R., and DISHION, T. (1983) "Early predictors of male delinquency: A review." *Psychological Bulletin* 94: 68-99.

MADDALA, G. S. (1983) *Limited-Dependent and Qualitative Variables in Econometrics*. Cambridge, UK: Cambridge University Press.

MIECZKOWSKI, T. (1990) "The accuracy of self-reported drug use: An evaluation and analysis of new data." In R. Weisheit (Ed.), *Drugs, Crime, and the Criminal Justice System* (pp. 275-302). Cincinnati: Anderson.

NORUSIS, M. J. (1990) *SPSS Advanced Statistics User's Guide*. Chicago: SPSS, Inc.

OHLIN, L. E., and DUNCAN, O. D. (1949) "The efficiency of prediction in criminology." *American Journal of Sociology* 54: 441-451.

SAS Institute, Inc. (1989) *SAS/STAT User's Guide* (Version 6, 4th edition, Volumes 1 and 2). Cary, NC: Author.

SCHAEFER, R. L. (1986) "Alternative estimators in logistic regression when the data are collinear." *Journal of Statistical Computation and Simulation* 25: 75-91.

SCHROEDER, L. D., SJOQUIST, D. L., and STEPHAN, P. E. (1986) *Understanding Regression Analysis: An Introductory Guide*. Sage University Paper series on Quantitative Applications in the Social Sciences, 07-057. Beverly Hills, CA: Sage.

SPSS, Inc. (1991) *SPSS Statistical Algorithms* (2nd edition). Chicago: Author.

STUDENMUND, A. H., and CASSIDY, H. J. (1987) *Using Econometrics: A Practical Guide*. Boston: Little, Brown.

WIGGINS, J. S. (1973) *Personality and Prediction: Principles of Personality Assessment*. Reading, MA: Addison-Wesley.

WOFFORD, S., ELLIOTT, D. S., and MENARD, S. (1994) "Continuities in marital violence." *Journal of Family Violence* 9: 195-225.

ABOUT THE AUTHOR

SCOTT MENARD is a research associate in the Institute of Behavioral Science at the University of Colorado, Boulder. He received his A.B. at Cornell University and his Ph.D. at the University of Colorado, both in sociology. His primary substantive interests are in the longitudinal study of drug use and other forms of illegal behavior. Dr. Menard's publications include the Sage QASS monograph *Longitudinal Research* (1991) and the books *Multiple Problem Youth* (with Delbert S. Elliott and David Huizinga, 1989) and *Juvenile Gangs* (with Herbert C. Covey and Robert J. Franzese, 1992).